Advances in Industrial Control

Springer
London
Berlin
Heidelberg
New York
Barcelona
Budapest
Hong Kong
Milan
Paris
Santa Clara
Singapore
Tokyo

Other titles published in this Series:

Identification of Multivariable Industrial Process for Simulation, Diagnosis and Control
Yucai Zhu and Ton Backx

Nonlinear Process Control: Applications of Generic Model Control
Edited by Peter L. Lee

Microcomputer-Based Adaptive Control Applied to Thyristor-Driven D-C Motors
Ulrich Keuchel and Richard M. Stephan

Expert Aided Control System Design
Colin Tebbutt

Modeling and Advanced Control for Process Industries, Applications to Paper Making Processes
Ming Rao, Qijun Xia and Yiquan Ying

Modelling and Simulation of Power Generation Plants
A.W. Ordys, A.W. Pike, M.A. Johnson, R.M. Katebi and M.J. Grimble

Model Predictive Control in the Process Industry
E.F. Camacho and C. Bordons

H_∞ Aerospace Control Design: A VSTOL Flight Application
R.A. Hyde

Neural Network Engineering in Dynamic Control Systems
Edited by Kenneth Hunt, George Irwin and Kevin Warwick

Neuro-Control and its Applications
Sigeru Omatu, Marzuki Khalid and Rubiyah Yusof

Energy Efficient Train Control
P.G. Howlett and P.J. Pudney

Hierarchical Power Systems Control: Its Value in a Changing Industry
Marija D. Ilic and Shell Liu

System Identification and Robust Control
Steen Tøffner-Clausen

Genetic Algorithms for Control and Signal Processing
K.F. Man, K.S. Tang, S. Kwong and W.A. Halang

Advanced Control of Solar Plants
E.F. Camacho, M. Berenguel and F.R. Rubio

Control of Modern Integrated Power Systems
E. Mariani and S.S. Murthy

Advanced Load Dispatch for Power Systems: Principles, Practices and Economies
E. Mariani and S.S. Murthy

Björn Sohlberg

Supervision and Control for Industrial Processes

Using Grey Box Models, Predictive Control and Fault Detection Methods

With 79 Figures

Springer

Björn Sohlberg, PhD
Dalarna University College, S-781 88 Borlänge, Sweden

ISBN 3-540-76221-3 Springer-Verlag Berlin Heidelberg New York

British Library Cataloguing in Publication Data
Sohlberg, B.
Supervision and control for industrial processes : using grey box models, predictive control and fault detection methods. - (Advances in industrial control)
1. Process control
I. Title
670.4'27
ISBN 3540762213

Library of Congress Cataloging-in-Publication Data
Sohlberg, B (Björn), 1949-
Supervision and control for industrial processes : using grey box models, predictive control and fault detection methods / B. Sohlberg.
p. cm. -- (Advances in industrial control)
Includes bibliographical references and index.
ISBN 3-540-76221-3 (alk. paper)
1. Process control. 2. Predictive control. 3. Fault location (Engineering) I. Title. II. Series.
TS156.8.S635 1997 97-31961
670.42'75--dc21 CIP

Printed in Great Britain

Typesetting: Camera ready by author
Printed and bound at the Athenæum Press Ltd., Gateshead, Tyne & Wear
69/3830-543210 Printed on acid-free paper

Advances in Industrial Control

United Kingdom
Professor Dr -Ing M. Thoma
Westermannweg 7
D-30419 Hannover
Germany

Professor H. Kimura
Department of Mathematical Engineering and Information Physics
Faculty of Engineering
The University of Tokyo
7-3-1 Hongo
Bunkyo Ku
Tokyo 113
Japan

Professor A.J. Laub
Department of Electrical and Computer Engineering
University of California
Santa Barbara
California 93106
United States of America

Professor J.B. Moore
Department of Systems Engineering
The Australian National University
Research School of Physical Sciences
GPO Box 4
Canberra
ACT 2601
Australia

Dr M.K. Masten
Texas Instruments
2309 Northcrest
Plano
TX 75075
United States of America

Professor Ton Backx
AspenTech Europe B.V.
De Waal 32
NL-5684 PH Best
The Netherlands

SERIES EDITORS' FOREWORD

The series *Advances in Industrial Control* aims to report and encourage technology transfer in control engineering. The rapid development of control technology impacts all areas of the control discipline. New theory, new controllers, actuators, sensors, new industrial processes, computer methods, new applications, new philosophies..., new challenges. Much of this development work resides in industrial reports, feasibility study papers and the reports of advanced collaborative projects. The series offers an opportunity for researchers to present an extended exposition of such new work in all aspects of industrial control for wider and rapid dissemination.

The steel industry world-wide is highly competitive and there is significant research in progress to ensure competitive success prevails in the various companies. From an engineering viewpoint, this means the use of increasingly sophisticated techniques and state-of-the-art theory to optimise process throughput and deliver ever more exacting dimensional and material property specifications. Dr. Björn Sohlberg's monograph demonstrates this interplay between fundamental control engineering science and the demands of a particular applications project in the steel strip production business. It is an excellent piece of work which clearly shows how these industrial engineering challenges can be formulated and solved.

The monograph has a number of outstanding features that the reader may find particularly interesting. Firstly, the whole process of constructing a grey box model for an industrial process is described in considerable clarity. The presentation covers the elements of the basic theory, and the steps in the application of this theory to a steel strip rinse process. The way in which the model development evolves to reach its final augmented extended Kalman filter form is particularly illuminating. Secondly, the monograph presents work on an optimised loss function approach for the design of a feedforward and feedback controller for the process. This work is related to the concepts of general predictive control. Finally, a useful contribution to fault detection and isolation at the process supervisory level is given. Again, the text covers the fundamentals, assesses the potential options and closes on an application to the rinse process. The Series Editors concur with Dr. Sohlberg's contention that many of the economic benefits are achieved by a scientific approach to supervisory level control and are aware of considerable industrial interest in this expanding technical area.

M.J. Grimble and M.A. Johnson
Industrial Control Centre
Glasgow, Scotland, UK

PREFACE

Many industrial processes are influenced by changing production parameters, for example within the steel industry, strip width and strip thickness. These parameters depend on the actual production and are measurable but cannot be used as variables to control the process. A change in a production parameter will cause a change in the output of the process. If the process is slow it will be a long time before a change will be detected and a long time before a feedback controller has compensated for the variation in the output.

In addition, many industrial processes have parts which wear out while the process is running. This means that the efficiency or the performance of the process is time varying. During planned stops for maintenance, the process is manually inspected and an operator has to decide whether a part should be replaced or not. Normally, a decision is taken based on experience or practice. If the parts subject to wear are replaced too often it gives rise to extra costs for spare parts and loss of production, and if the parts are replaced too seldom, the process will not behave properly.

One of the goal of this text is to illustrate the application of control theory and methodology in a real industrial plant for to the problem discussed above. The objectives include both control and supervision of a process within the steel industry. The book aims to bridge the gap between theory and practice, and furthermore, encourage academic researchers and industrial practitioners to cooperate.

The first part of the book deals with grey box modelling. It is a promising method to make use of the existing partial knowledge of an industrial process and develop the model further by using measured data from the process. The procedure goes on until the goal with the model is fulfilled. This modelling method is applied to the process under study. The rest of the book deals with model-based control and supervision with application to the same process. Two different type of controllers are discussed and tested in the plant: a combined feedforward and feedback controller, and a model predictive controller. For the supervision of the process we distinguish between two kind of faults: slowly changing and abrupt changing faults. The first often originate from machine wear and the latter from abrupt errors within the process control system. In practice, a supervision system has to deal simultaneously with both types of faults, but here we make a distinction because different methods have to be used.

Acknowledgements

The author would like to thank a number of people who made this monograph possible. At the Royal Institute of Technology, I would like to thank Professors T. Bohlin and B. Wahlberg, for many valuable comments and discussion. At the SSAB Tunnplåt AB, I acknowledge; the production managers, the engineers who made the installation of electrical equipment and the process operators, for enthusiastic collaboration during the experiments with the process. My thanks to Dr M. Nihtilä, University of Kuopio, Dr M. A. Johnson, University of Strathclyde and Dr. E. W. Jacobsen, Royal Institute of Technology, Stockholm for comments and suggestions on the manuscript. Finally, M. Hammar, for proof-reading and correcting the English grammar.

To write a book takes up one's leisure time to a great extent. Therefore, my thanks to my family Gertie and Jacob, who have shown great patience and understanding.

Björn Sohlberg
Centre for Industrial Engineering and Development
Dalarna University College, Sweden

GLOSSARY

Table 1 Variables in the model structure of the case study.

Name	Type	Meaning	Range	Unit
C_1	State variable	Concentration	0 - 10	kg/m^3
C_2	-"-	-"-	0 - 1	-"-
C_3	-"-	-"-	0 - 0.1	-"-
C_4	-"-	-"-	0 - 0.02	-"-
C_5	-"-	-"-	0 - 0.005	-"-
F_{c1}	Control input	Flow into tank 5	0 - 6	m^3/h
F_{c2}	-"-	Flow into tank 2	0 - 1	-"-
B_b	Extra input	Strip width	600 - 1800	mm
B_v	-"-	Strip velocity	0 - 4	m/s
B_t	-"-	Strip thickness	2 - 7	mm
L_1	Output	Conductivity	0 - 10000	mS/m
L_2	-"-	-"-	0 - 1000	-"-
L_3	-"-	-"-	0 - 100	-"-
L_4	-"-	-"-	0 - 20	-"-
L_5	-"-	-"-	0 - 5	-"-
$F_{b1} \ldots F_{b6}$	Intern. variable	Flow via strip	Varying	m^3/h
$F_{a1} \ldots F_{a5}$	Intern. variable	Evaporation	Less then 0.04	m^3/h

Table 2 Known constants in the model structure of the case study.

Name	Type	Meaning	Value	Unit
V_1	Constant	Volume tank 1	6.0	m^3
V_2	-"-	Volume tank 2	5.1	m^3
V_3	-"-	Volume tank 3	6.0	m^3
V_4	-"-	Volume tank 4	6.8	m^3
V_5	-"-	Volume tank 5	9.9	m^3
A_1	-"-	Bath area tank 1	8.8	m^2
A_2	-"-	Bath area tank 2	8.8	m^2
A_3	-"-	Bath area tank 3	8.8	m^2
A_4	-"-	Bath area tank 4	8.8	m^2
A_5	-"-	Bath area tank 5	11.2	m^2
R_c	-"-	Radius	0.1	m
L_m	-"-	Transducer scale	1100	$mS \cdot m^2/kg$
b	-"-	Slit width	2.0	m
F_s	Presumed constant	Circulated flow	100	m^3/h
F_{m1}	-"-	Flow from tank 1	0.3	m^3/h
C_0	-"-	Concentration	200	kg/m^3

Table 3 List of symbols.

Symbol	Meaning
$B_{toff1} \dots B_{toff5}$	Unknown parameters
E	Expectation symbol
F	General discrete process model
G	General discrete output model
H	Hessian matrix
J	Performance measure
K	Kalman filter gain
$K_{b1} \dots K_{b5}$	Unknown parameters
$K_{t1} \dots K_{t5}$	Unknown parameters
L	Feedback control matrix
M	Fisher information matrix

Table 3 (Continued) List of symbols.

Symbol	Meaning
N	The number of samples
P	Covariance matrix
Q_0 Q_1 Q_2	Weighting matrices
R(k)	Covariance matrix of the innovation process
R_1	Covariance matrix, process noise
R_2	Covariance matrix, measurement noise
S	Sensitivity function
V	Scalar loss function
$\overline{F}$	General augmented discrete process model
$\overline{G}$	General augmented discrete output model
$\mathcal{H}$	Hypothesis
$\mathcal{h}$	Control function
$\mathcal{L}$	Likelihood function
$\mathcal{M}$	Model structure
$\mathcal{N}$	Weighted sum of squared residuals
e(k)	Innovation process
f	General continuous process model
g	General continuous output model
n	The number of states
p	The number of outputs
r	The number of inputs
$r_{ee}(\tau)$	Auto-correlation function
$r_{eu}(\tau)$	Cross-correlation function
s(k)	Auxiliary input vector
u(t), u(k)	General input signal
v(k)	Process noise
w(k)	Measurement noise
x(t), x(k)	General state variable
y(t), y(k)	general output signal
ϕ	Regressor vector
η	Reduction in concentration between nearby tanks
θ	Unknown parameter vector
ξ	Augmented state vector
Γ	Linear measurement matrix
Φ	Linear transition matrix

Table 4 List of abbreviations.

Abbreviations	Meaning
AEKF	Augmented Extended Kalman Filter
AIC	Akaike's Information Criterion
ARMAX	Autoregressive Moving Average Exogenous input
det	Determinant of a matrix
diag	Diagonal elements of a matrix
dim	Dimension of a matrix or a vector
EKF	Extended Kalman Filter
GLR	General Likelihood Test
KF	Kalman Filter
LQ	Linear Quadratic
LQG	Linear Quadratic Gaussian
MPC	Model Predictive Controller
MTBF	Mean Time Between Failure
MTTF	Mean Time To Failure
MTTFF	Mean Time To First Failure
MTTR	Mean Time To Repair
PC	Personal Computer
PRBS	Pseudo Random Binary Signal
US$	United States Dollar
WSSR	Weighted Sum of Squared Residuals

CONTENTS

1 BACKGROUND

1.1 Introduction

Competition within the steel industry has increased during the last decade. One consequence is that the product quality has become very important when marketing steel products. Since the steel industry is an energy demanding industry and the costs of energy have increased, it is necessary to reduce the consumption of energy. Furthermore, there is an increased demand for new properties of iron and new dimensions. As a consequence, the number of reference adjustments has increased when controlling an industrial process. To fulfil these requirements, novel control methods and new processes have become very important to produce products with a competitive quality and price.

Proper functioning of the processes has also become very important for preventive and predictive maintenance. At a higher control level, diagnosis and supervision of industrial processes have become elements of an operator guidance system. For that purpose, we can use a model of the process to better understand the behaviour of the system. Specifically, at a higher control level, the process model can be used to:

- Detects faults of actuators, process components and sensors.
- Diagnose the size and location of a fault.
- Determine whether it is possible to proceed to run the process although the system is not functioning properly.
- Predict the influence of long term wear.

The supervision purposes of industrial processes are to advise the operator about what is happening and what to do about it. For industrial applications, it is not sufficient to make a certain diagnosis, but it is also important to quantify the probability that an alternative diagnosis is correct. With this information, together with the economic consequences of certain decisions, optimal decisions can be made.

1.2 Rinsing process

As a case study for this work, we use a rinsing process within the steel industry. This process is a part of the pickling line at Domnarvet Steel plant, owned by the Swedish Steel Corporation, SSAB. After the steel strip has passed a bath of hydrochloric acid, the steel strip is rinsed by passing through a rinsing process. The rinsing process consists of several connected rinse tanks. There are similar processes in various other parts of the steel industry. In addition, there are processes based on the same principle in paper mills and the food industry.

The purpose of the rinsing process is to achieve a well-rinsed steel strip that is free from contamination. This is obtained by allowing the steel strip to pass the rinse tanks continuously. The rinse tanks contain increasingly cleaner rinse water. The purity of the last rinse tank is crucial for the purity of the steel strip. By feeding clean water into the last rinse tank, the contamination in this tank can be limited and be maintained below a given reference value.

Because of several production variables, the amount of acid transferred from the acid tanks into rinse tanks varies. This means that the purity of the rinse water, and indirectly the cleanness of the steel strip depend on the actual production.

Two of the variables, the strip width and the strip thickness, are results of planning the production, while one of them, the strip velocity, depends on other technical circumstances that have to do with production. The production variables are measurable when the process is running. The rinsing process is also influenced by other disturbances that are not measurable.

The dynamics of the process are characterised by long time constants. A consequence is that it takes a long time to correct high acid in the rinse tanks. The process is also non-linear, since the time constants and the gain for the process depend on the production variables and the inflow of clean water. The plant is also characterised by the fact that it consists of several subprocesses, connected by different flows and depending on the acidity of the rinse water in each subprocess. The interdependence between the different subprocesses is complicated and difficult to grasp intuitively.

As is to be expected, certain components of the process are subject to wear when the process is in operation. This means that the function or efficiency of the process deteriorates with time. To achieve a well-working process, it is necessary that the parts subject to wear have an acceptable condition or efficiency. During planned stops for maintenance, the process is manually inspected by an operator and worn out parts are replaced with renovated spare parts. It is often difficult to judge by manual inspection which parts are in the worst condition and to decide when to replace a worn out part.

In the case of the rinsing process, worn parts are not simply replaced with new ones. A key element for the process is the renovation of worn parts. Renovated spare parts have varying performances. The performance depends on the quality of the spare parts and how many times a spare part has been renovated. This means that it is not possible to know the condition of a spare part in advance.

The technology of the rinsing process has been treated earlier by Kushner (1976). Stein (1988) described an application of a rinsing process, which consisted of several connected rinse tanks. In this application, modelling of the stationary states in the rinse tanks is treated. This work is also based on the assumption that the flow via the steel strip is equal at the beginning of every rinse zone. In addition to these works, rinsing is treated by Mohler (1977) and Ledding (1986). These articles describe applications of rinsing processes based on equations presented by Kushner.

There are several reasons for applying novel control theory to this rinsing process:

- The process consists of several coupled sub-systems, making it difficult to assess the influence of the control signal. Accordingly, the control algorithm can be improved by constructing a controller based on a model of the process.
- The process is influenced by disturbances which are unknown and not possible to measure. For the rinsing process, there is no compensation for this type of disturbance. Consequently, the mean value of the cleanness of the rinse water must be kept unnecessarily high. Otherwise, the disturbances will result in an incompletely washed steel strip.
- The efficiency of the process is also influenced by the wear. There is no compensation for the degree of wear of the parts subject to wear. As a consequence, the efficiency of the process becomes a function of the deterioration.
- Another reason for applying advanced theory on the actual process is to supervise the parts of the process that wear out when the process is running. Supervision of these parts demands a model of the process and on-line measurement data. To distinguish between the effects of worn parts and non-linearities of the process, it is necessary to build in the non-linearities in the model and model the wear explicitly. With this type of model, eventually, it is possible to estimate unknown parameters modelling the worn parts of the process.

To solve these problems, a model of the process is developed which is based on systematic use of a priori knowledge of the process and data from experiment. This method is called grey box modelling. The process is divided in several sub-processes and parts subject to wear are modelled explicitly. It makes the model suitable to compensate for auxiliary production parameters and supervise the condition and faults of the process.

1.3 Outline of the chapters

The purpose of this work is to study application of control theory and methodology in a real industrial plant in the field of model-based process control, specially focussed on a process within the steel industry. The disposition of this book is as follows:

In chapter 2, we introduce the concept of grey box modelling. The method is distinguished from black box identification and white box modelling. Black box identification needs only identification and white box modelling requires only modelling, but grey box modelling needs both modelling and identification. Grey box modelling consists of several procedures such as; basic modelling, expanding modelling, experimentation, parameter identification, model analysis and model appraisal.

Chapter 3 focusses on a case study of basic modelling. In chapter 4, we proceed to expand the model to describe the process more precisely. The purpose of the model is considered when we decide to stop expanding the model and when we consider the model good enough. To make supervision of the wearing parts possible, these parts are modelled explicitly. It must also be possible to use the model to estimate unknown parameters on-line. The model will also be used as a basis to construct a control algorithm and to compute a control signal.

Chapter 5 describes on-line estimation of unknown parameters. An Augmented Extended Kalman Filter is used to achieve the estimates. In chapter 6, a combined feedforward and feedback controller is first discussed. It forms part of the family of gain scheduling controllers. The feedforward part compensates for variation of the production variables. The feedback controller compensates for other disturbances influencing the process, and thus compensates for the fact that the feedforward control cannot be perfectly made. The controller is in a way heuristic, because we want the controller to be easy to implement. Secondly, an optimal controller is discussed. This controller is based on a loss function, which is minimised at every sample. This type of controller is closely related to a general predictive controller. The two types of controllers are first simulated with a model of the process and secondly tested with the real process.

Chapter 7 deals with supervision of wearing and a rule base for decision to replace worn parts is established when these parts are to be replaced by renovated parts. The supervision of the process is based on the grey box model developed in Chapter 3 and 4. On-line estimation of unknown parameters is based on the method presented in Chapter 5. The rule base gives guidance in deciding which worn parts should be replaced at the next planned stop for maintenance.

Chapter 8 treats fast changing faults. Different fault detection methods are given such as the Weighted Sum of Squared Residuals and the General Likelihood Ratio methods. A effective method for both fault detection and diagnosis is to use a bank of Kalman filters. Both advantages and disadvantages are discussed. The chapter ends with an application to the rinsing process by using the Weighted Sum of Squared Residuals method.

2 GREY BOX MODELLING

2.1 Introduction

In this chapter the concept of grey box modelling is introduced. The method is distinguished from conventional methods of making models of industrial processes. We will also discuss the purpose of grey box modelling and when the method ought to be used.

Conventionally, a process model is based on either complete knowledge of the process, named "white-box" model or based principally on experimental data, a "black box" model. The white box model is also referred to as simulation model. The difference between these modelling principles and grey box modelling, is the systematic use of partial a priori knowledge of the process and data from an experiment.

The construction of a grey box models can be divided into several activities such as; basic modelling, expanding modelling, experimentation, identification, model analysis and model appraisal. First, we try to find a basic model, which is used as a basis for further expansion. Expanded modelling means improving the basic model by using knowledge created from model analysis and other possible sources. To perform the model analysis and model appraisal, the role of a human designer is stressed. It depends on the skill of the model builder and how the results from the model analysis procedure are evaluated.

For successful development of a process model, we need efficient software for identification. The software must be able to handle non-linear models and consist of a set of tools for model analysis.

This chapter starts with the justification for applying grey box modelling for industrial processes. Specially, the purposes of the model is considered. We discuss also the possible degree of greyness of different kinds of processes. The main effort in this chapter is focussed on a construction procedure of grey box models.

2.2 Background for grey box modelling

When constructing a mathematical model of a given physical process, two different ways of tackling the problem can be used, (Bohlin, 1991a). The first way is based on the presumption that the process can be described completely by mathematical equations, such as differential equations, algebraic equations, logical relationships and similar types of equations. This type of model is called a white box model. White box models can be found in the field of electrical net theory. By using Kirchoff's law and similar theorems, a mathematical model of an electrical circuit is constructed. Some unknown parameter may have to be estimated from a simple experiment or by fitting.

The second way is based on the idea that the process is regarded as completely unknown and that it is not necessary to use any model structure which reflects the physical structure of the process. This type of model is described as a black box model, since a model from a group of standard models is used to describe the process. The unknown parameters in the model are estimated using data from an experiment with the process and the model gives an input/output relation of the process. Examples of processes modelled with black box models are economic, social and ecological systems. This is a class of processes, whose underlying causes are complicated and uncertain. Black box models are also common to model industrial processes, especially in connection with adaptive control.

For many industrial processes there exists some but incomplete knowledge about the structure of the process. The amount of knowledge varies from one process to another. Between white box models and black box models, there is a grey zone. There are processes about which there is some, but incomplete knowledge about the structure of the process, see Fig. 2.1, (Karplus 1976). The cases between white box and black box are called grey box modelling.

White box model design needs only modelling and black box design needs only identification. However, grey box design requires both modelling and identification, since the result from modelling is not certain and must be verified by identification, (Bohlin 1991a).

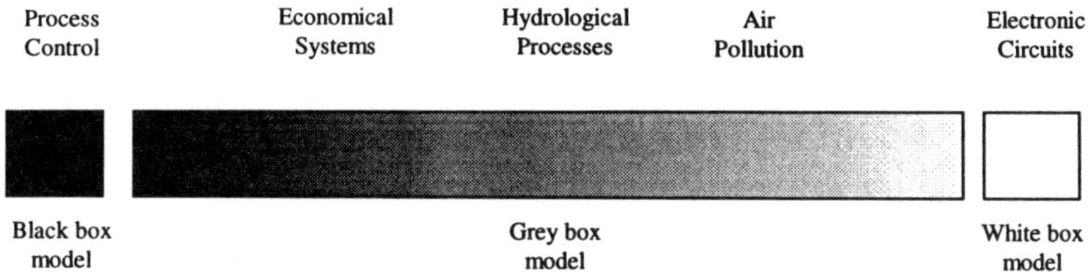

Fig. 2.1 Grey scale for grey box modelling.

The advantage of a grey box model compared to a black box model, is the possibility to achieve a better model of the process. For example, it is possible to build in non-linearities in the model and separate different parts of the process into submodels. This may be necessary if the model is to be used to supervise some parts liable to wear.

Within the steel industry, there are measurable production variables which influence the behaviour of the process. For example strip velocity, strip width and strip thickness. With a good model of the process it is possible using a feedforward controller or other more sophisticated controllers, to compensate for variations in these production variables.

Since it may be possible to divide the process model into submodels or internal models, the model achieved with grey box modelling can be used to optimise the performance of the system. It may be possible to rearrange the interconnection of different process parts in a more efficient way. For example, the control signals should be connected to another subprocess to make better use of the control action.

As mentioned earlier in this chapter, grey box modelling includes both modelling and identification. The model is based on available physical knowledge. The knowledge is not only incomplete but also uncertain. The data which are used to identify the unknown and uncertain parts are also influenced by unknown disturbances. Consequently, the resulting model has to be analysed by using measured data. The identification and analysing procedure may give more information on how the unknown and uncertain parts should be designed. This gives an opportunity to achieve more knowledge about the functionality and physical understanding of the process. As a result the process can be controlled more efficiently, and the quality of the product produced can be improved.

The grey box modelling method may also be appealing when the process is non-linear. In these cases it may be difficult to achieve a satisfactory black box model when the operating point is changing, and it may be necessary to use several black box models, which are related to the operating points.

It must be pointed out that a grey box model may be more expensive to develop than a black box model, since it may takes longer to construct a grey box model than a black box model. However, it is the purpose of the model that decides what kind of model we need and whether it may be profitable to use a grey box model.

2.3 Model construction procedure

The construction procedure of a grey box model can be divided into different sub-procedures. Here, they are presented as separate activities, but in practice the activities are closely related to each other.

- **Basic modelling.** This means setting up a basic structure of the model based on a priori knowledge about the process. The physical knowledge of the process is presented with a set of equations. These can be different types such as; non-linear differential equations, algebraic equations or other types of expressions.
- **Expanded modelling**. The basic model is expanded to model unknown and uncertain parts of the process. This is often an engineering task. Great demands are put on the model builder to use the information from the model analysis and model appraisal. He or she must also have an intuitive feeling of how to expand basic models.
- **Experiment**. For the identification of the unknown parts, it is necessary to have measured data from the process. The data has to be "informative". This means that the measured signals must vary enough due to variation in the input signals, so the data can be used for identification. The basic model might give some information on how to perform the experiment.
- **Identification**. The tentative model consists of unknown parameters. These are estimated by using a software which can handle non-linear models, since models based on a priori knowledge are seldom linear.
- **Model analysis**. The resulting model is analysed by the following tools; simulation, statistical methods, the value of the loss function, cross validation of the estimated parameters and reliability of the estimated parameters.
- **Model appraisal.** The guiding star for the work to produce a process model, is the purpose of the model, since the model is going to be used in some way, for instance: feedback control, feedforward control, model predictive control supervision or failure detection. Different model purposes demand different kinds of models. The results from the model analysis are used to appraise the resulting model and give hints for further model expansion.

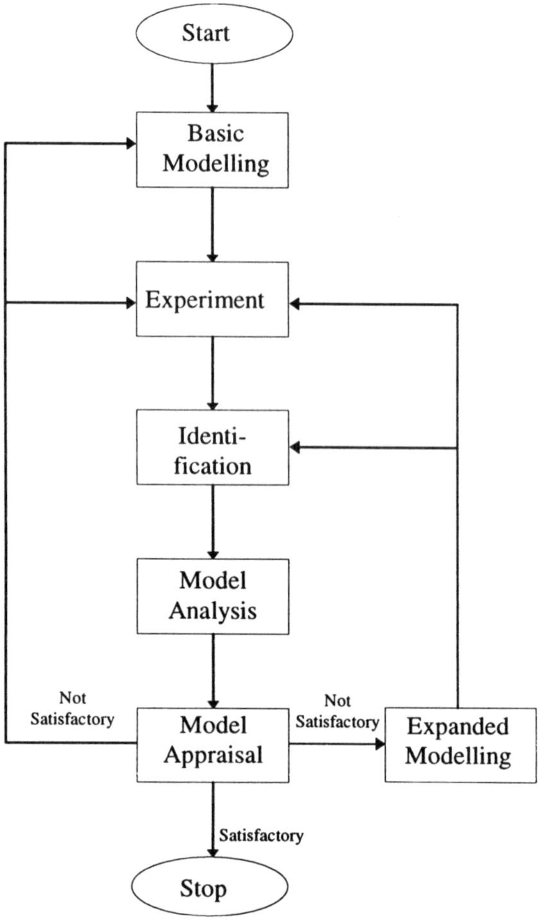

Fig. 2.2 Flow chart.

A flow chart of the model construction procedure is shown in Fig. 2.2. The basic model is complemented with components describing the unknown parts of the process. After identification, the resulting model is analysed using different methods. It is then necessary to achieve as much information as possible about the process. By means of the model appraisal we decide whether the model fulfil the purpose of the model or not.

When the model does not meet the purpose, there might be information from the model analysis which can be used to improve the model. Otherwise, the model has to be expanded by using other assumptions of the process or we need to change the basic model. Another possibility is to make a new experiment based on different sequence of input signals or different circumstances.

The procedure described in Fig. 2.2 consists of six separate blocks, which are performed in a sequence. However, in the real situation, work will not proceed in such a well structured way, since the model construction procedure involves a model builder who ought to play an active roll during the operation, from defining the purpose of the model to producing a useful model. Therefore, the scheme in Fig. 2.2 should be extended to include a human designer. Furthermore, interaction between the blocks is achieved by and through the user. This means that we sketch the modelling procedure with the designer in the centre, who is surrounded by the different activities, see Fig. 2.3. The priority of the different activities is now managed by the model builder.

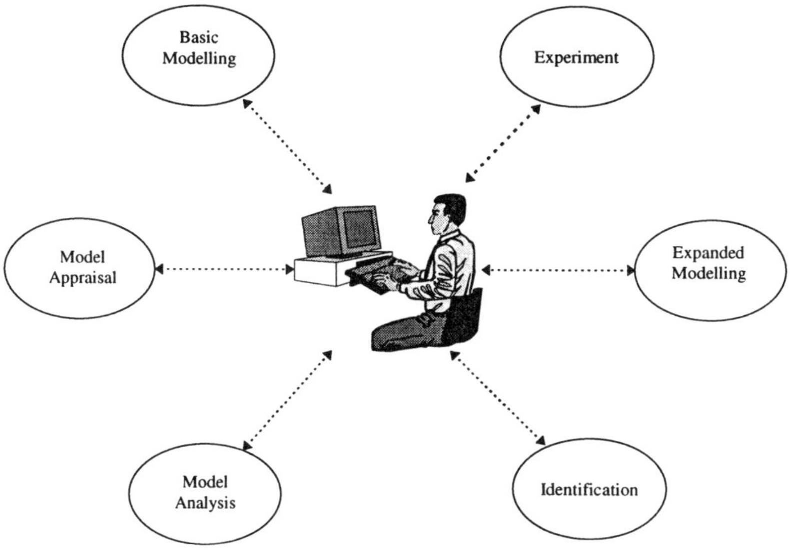

Fig. 2.3 Human designer interaction.

2.3.1 Basic modelling

In the classic approach to mathematical modelling it is assumed that the a priori information is exact and is based on scientific theory and be can presented by a set of well defined equations. Modelling of physical systems are treated by Ljung and Glad (1994), Ordys et. alt (1994), Rao et. alt (1993), and Spriet and Vansteenkiste (1982).

Grey box modelling also needs a basic structure of the process. The goal of basic modelling is to describe the relevant knowledge and form a basic model of the process. This is in practice not an obvious task, because one cannot be sure what is real facts and what is supposed to be facts about the process considered. This raises the question if it is at all possible to form a proper basic model to start from. However, this philosophic discussion is beyond the scope of this book. Despite the uncertainty the basic model should be constructed systematically. We distinguish between the following steps during the basic modelling procedure:

Step 1. Process description.

This means that the function of the process and data about the process are literally formulated. The process description should include the purpose of the process. The description should also contain schematic figures of the process.

Step 2: Process structure.

When the problem is structured, the process is divided into subprocesses and we describe what kind of connections there are between the sub-systems. At this stage, the input and the output signals are also listed. The input signals need to be divided into signals possible to control and signals which are only possible to measure. It may be necessary to perform the procedure of dividing the system into subprocesses at several stages, so the function of the different parts is as clear as possible.

Step 3: Process hypotheses.

Several hypotheses $\{ \mathcal{H}_i : i=1...v \}$ must be formulated about the behaviour of the process, before the system can be described using mathematical equations. The hypotheses have to be analysed to make sure that they are reliable.

Step 4: Basic equations.

Based on the subsystems and the process hypotheses, we can formulate the relations between the variables for the process. As mentioned earlier, physical insight is used to form the basic equations. Typically, the equations consist of three types: balance equations, algebraic equations and logical expressions.

I Balance equations.

The balance equations relate dependent variables to the independent variables and are derived from the principle of conservation, (Stephanopolus, 1984), which is formulated as:

$$\frac{\text{Accumulation of Q}}{\text{Time period}} = \frac{\text{Total input of Q}}{\text{Time period}} - \frac{\text{Total output of Q}}{\text{Time period}}$$

$$+\frac{\text{Total generation of Q}}{\text{Time period}} - \frac{\text{Total consumption of Q}}{\text{Time period}} \qquad (2.1)$$

where Q denotes:

- Energy
- Mass
- Momentum

The equation (2.1) means that the accumulation of a quantity Q per time period within the process is equal to the net sum of total input to the process plus the net sum of Q generated inside the process.

II Algebraic equations.

An algebraic equation gives a static relation between process variables. The relations can be linear or non-linear. An example of an algebraic equation is Ohms law, the relation between current and voltage related to a resistor, see Fig. 2.4. Ohm's law gives an algebraic relationship as $u = R \cdot i$.

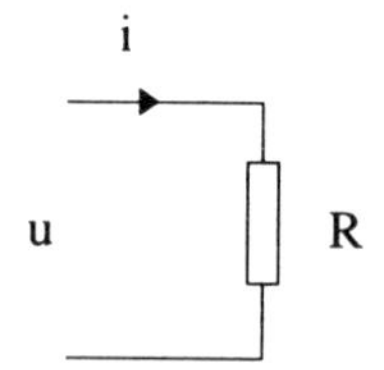

Fig. 2.4 Algebraic relation.

III Logical expressions.

There is a class of non-linearities which must be expressed by a logical expression. A typical relation of this kind is shown in Fig. 2.5, which is a relay. For example the relay can be described with the following logical expression:

IF $\{E > 0\}$ THEN $\{U = +1\}$ ELSE $\{U = -1\}$

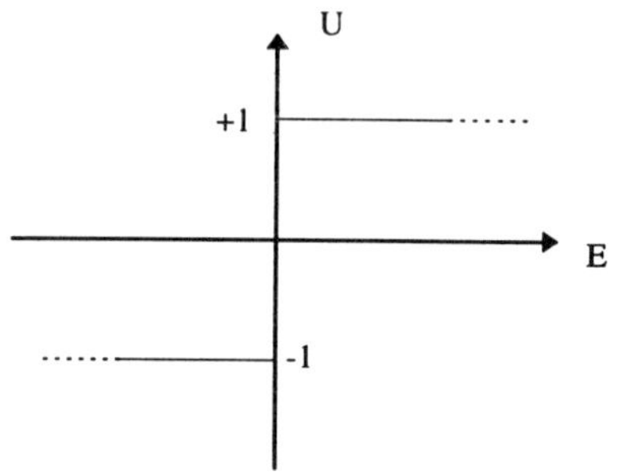

Fig. 2.5 Logical relation.

Step 5: State space model.

The equations formulated in step 4, construction of the basic equations, give a core description of the process. However, the equations need to be structured so they can be used for identification. This is done using the following procedure:

(i) Choose a set of state variables $\{x_1\ x_2\ x_3\ \ldots.\ x_n\}$. This is done from a physical point of view.

(ii) The derivatives of the state variables are expressed as functions of the states and the inputs $\{u_1\ u_2\ u_3\ \ldots\ u_r\}$.

(iii) Incorporate the algebraic equations which relate different kinds of variables.

(iv) Complete the model description with prospective logical expressions.

(v) Connect the output variables, $\{y_1\ y_2\ y_3\ \ldots\ y_p\}$ to the states and also if required to the inputs.

Step 6: Model preparation.

(i) *Discretization.* Normally, the basic model is continuous and nonlinear. This is natural since the model is based on physical insight. The model consists of unknown parameters which have to be estimated using a suitable software, see Section 2.4. To reduce the time for searching for an unknown parameter it is a good idea to use a discrete version of the model. This will decrease the estimating time considerably. The identification software may also need a process model in discrete form. Under specific circumstances the discretization can be done easily using Euler's method, (Åström and Wittenmark, 1997):

$$\frac{dx(t)}{dt} = \frac{x(t+h) - x(t)}{h} \qquad (2.2)$$

Note, Euler's method is not a suitable method when we are dealing with stiff system. In such cases, the sample time is very crucial and it is better to solve the problem by using a identification software which handle both integration and identificatin, see section 2.4.

(ii) *Scaling.* When the measure range of the process signals is relatively different, scaling may be necessary to achieve success during the identification. Scaling is, then, fundamental for process inputs and outputs. The scale factors are not crucial, a rule of thumb is to use the mean value of the data sequence for a specific signal

(iii) *Process noise.* A model structure of a real process involves many simplifications and uncertain assumptions. In practice, it is impossible to make the perfect model. We have to accept that there are parts of the process which are not possible to describe or have to be modelled with some ad-hoc term. Apart from the unmodelled parts there are disturbances acting on the process originating from other sources. Unmodelled parts of the process and other disturbances are added as disturbances to the states of the model. As a first approach, during identification these disturbances are regarded as white noise. This is also a simplification, but colouring the state noise needs extra states and prolongs the identification time. However, we have to be carefully because this simplification is only valid for fast varying disturbances. When the disturbances is a slowly varying process we need extra states to model the influence of the disturbances.

(iv) *Measurement noise.* The measurements are also corrupted by disturbances. There may be sources within the measurement devices, induced disturbances while transmitting the signal from the transducer to the datacollecting equipment and discretization errors when the analog signal is converted to a discrete form.

Step 7: General form.

For a discrete version of the model, the resulting process model consists of difference equations and algebraic functions relating the process states to outputs. The model is formalised as the following equations:

$$x(k+1) = F\left[x(k),\ u(k),\ \theta,\ k\right] + v(k) \qquad (2.3)$$

$$y(k) = G\left[x(k),\ u(k),\ \theta,\ k\right] + w(k) \qquad (2.4)$$

where x(k) is the state vector, y(k) the output vector, u(k) the control vector and θ(k) an unknown parameter vector. The disturbances are represented by v(k) and w(k), with are zero mean and covariance matrices R_1 and R_2.

2.3.2 Experiment

Theoretical results of experiment design for linear systems are given by Goodwin & Payne (1977) and Goodwin (1982). A performance measure relates the expected accuracy of the estimated parameters to a given model structure. To measure the information richness of an experiment, the Cramer-Rao lower bound is used. The lower limit of the parameter covariance matrix gives in equation(2.5).

$$\text{cov}\left(\hat{\theta}\right) \geq M^{-1} \tag{2.5}$$

$$M = E\left(\frac{\partial}{\partial\theta}\right)^{T}\left(\frac{\partial}{\partial\theta}\right)\mathcal{L}(y,\theta) \tag{2.6}$$

where M is the Fisher information matrix and $\hat{\theta}$ is the parameter estimate, y is the observation of a stochastic variable and $\mathcal{L}$ denotes the likelihood function.

A design procedure of optimal inputs for parameter identification of a non-linear system is developed by Goodwin (1971). This leads to minimum variance estimates of the parameters and the states in the model. A realistic assumption for this procedure is the constraints on the inputs and the states. The model can also be corrupted with both white and coloured noise. The optimal input will be chosen to minimise the following performance index:

$$J = \text{Trace}\left(M^{-1}\right) + \sum_{k=1}^{N} V\left\{x(k), u(k), k\right\} \tag{2.7}$$

where V is a scalar loss function, x(k) is a state vector u(k) is an input vector and k is a sample index.

One of the primary distinguishing features between linear and non-linear models is the amplitude dependence of the behaviour of non-linear models. In non-linear identification it may be necessary to consider the response of the process over a range of input amplitudes, (Henson and Seborg, 1997).

For grey box modelling, the data from an experiment are used both for parameter estimation and model testing. This means that the data are used also to develop the model further and it should be possible to discriminate between several models. The problem can be formulated as follows: if there are a number of models, each can be embedded in a more general model, (Atkinsson, 1975), and the design of an experiment can be based on the general model. Optimal input signals for parameter estimation in dynamic system is given by Mehra (1974). Several extensions to state restriction and non-linear system are discussed.

Experiment design criteria for applications on grey box modelling is given by Melgaard H. et. alt (1993). One criterion is that the process should be exposed with sufficient stimulation. This means that the experiment should excite at least all states of the process that can be affected when the model is to be used.

Another property of the experiment is the reproducibility, which is used for cross-validation. The experiment is divided into two data sequences; one sequence is used for identification and the other sequence is used for analysis of the model. The circumstances for the process should be the same for the two different data sequences. For instance, if the process is stopped for maintenance and some parts are replaced, the behaviour of the system may be affected. A similar experimental circumstance as reproducibility is that the experiment should be carried out on the same system as that on which the model is going to be applied. This alludes to the purpose of the model and Bohlin (1991) denotes it as a basic rule for experiment design.

Practical experimental design for processes within the steel industry is often troublesome. This is also the situation for similar types of systems within the process industry. Many considerations must be included and it is not obvious how to perform the experiment. The test has to be done in controllable circumstances. Consideration must be taken to run the process safely and also the quality of the material produced must be guaranteed. However, from the above we have a well-establish base and guidance for experiment design.

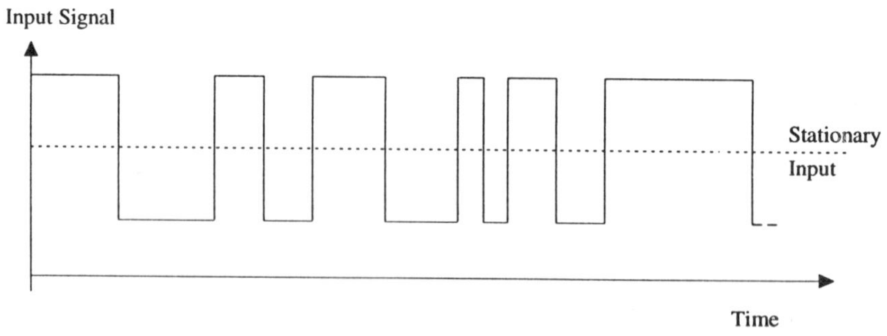

Fig. 2.6 Pseudo Random Binary Signal.

One of the goals for a realistic experiment is to have a sequence of input-output data which is "informative". This means that the signal/noise ratio of the output is "good" and the disturbances will play a smaller role. Futher, the input sequence must be uncomplicated to apply. A commonly use input signal is a pseudo random binary signal (PRBS), Fig. 2.6. Experiments using PRBS-signals is addressed by Johansson (1993), Ljung (1987), and Söderström and Stoica (1989).

❐ **Comments on the PRBS signal**

1. The stationary point of the input should be chosen as the value which is going to be used when the model is applied. When the model will be used for several stationary points it will be necessary to perform the experiment with either increased changes in the input signal or performed on the different stationary points. This also applies to other circumstances which will influence on the experiment.

2. There are also aspects on how much the amplitude of the input shall be varied. Small variations will not give informative outputs. However, large amplitudes may affect the process negatively. The amplitude is chosen as a trade-off between "informative" signals and proper behaviour of the process.

3. The duration of a change in amplitude has to be long enough to cause an effect on the output. On the other hand too long length will not give any more dynamic information about the system. To achieve as much information as possible, the period interval should vary. A reasonable choice is to use a rectangular statistical distribution with the shortest interval as 10% and longest interval as 90 % of the rise time of the process.

4. The data is normally picked up with a constant sampling interval. When the interval is too short, the noise will dominate the changes in the output. On the other hand, a long sampling interval will not detect the dynamics of the process. The choice of sampling interval is made in the light of informative change in the output between the sampling instances. A rule of thumb is: a suitable choice is about 10 - 20 % of the time constants of interest.

5. The number of samples must also be decided. This settles the duration of the experiment. When the number is increased, the variance of estimates will decrease. A rule of thumb is that the experiment length should be longer than ten times the dominating time constant, (Johansson, 1993).

6. A PRBS-signal is basically constructed to identify linear system. This type of input signal has low impact on the behaviour for low frequencies. However, we can compensate for that problem by prolong the switching interval.

2.3.3 Identification

The tentative model includes unknown parameters which have to be estimated. A grey box model, based on physical insight, often consists of non-linear parts. In the literature there are several methods for parameter identification. They all make use of the difference between the process and model one-step predicted output. The difference will form a time sequence, which is called the innovation process.

The measured data is the response from the process. We can suppose that the data is random, but it will also contain some representative process information. A common way to represent the information is by using a probabilistic approach. We make use of the Bayesian ideas and consider the estimate of the unknown parameter θ to be a random variable denoted as $\hat{\theta}$.

The fundamental of Bayesian statistics is the Bayes rule, (Box and Tiao, 1973), (Press, 1989) and (Sage, 1987). Identification is based on optimisation of the Likelihood function $\mathcal{L}$, given by equation (2.8), (Eykhoff, 1974).

$$\mathcal{L}=\frac{1}{2}\sum_{k=1}^{N}\left[e(k)^{T}R(k)^{-1}e(k)+\log[\det R(k)]\right]+\frac{1}{2}p\cdot N\cdot\log(2\pi) \tag{2.8}$$

where

p the number of outputs, $p=\dim[y(k)]$.

N the number of samples.

e(k) the prediction error between the process output and the model output, i.e. the innovation process.

R(k) the covariance matrix of the innovation process.

❐ **Remark**

The loss function is derived based on the assumption that the innovation process is white and normally distributed. This is an ideal situation, which in practice seldom happens. However, it focusses on the fact that the validation procedure should include whiteness test of the innovation process.

❐ **Remark**

Under the assumption of grey box modelling, the parameter vector θ has some physical meaning or explanation. This type of a priori knowledge is involved in the grey box modelling procedure since the estimate of θ is restricted to a defined region. It is used in at least three ways: first we need to initiate the parameters sought, secondly the convergence of the parameter estimation can be made more efficient if the estimate is restricted. Third, the estimates should belong to the defined region. If this is not the case, we have to decide what should be done about it.

❐ **Remark**

The loss function $\mathcal{L}$ is a function of the parameter vector θ. The identification procedure means to finding an estimate $\hat{\theta}$ of θ, which minimises $\mathcal{L}$. The problem is solved by parametric optimisation and iteratively finding a better estimate of θ through some numerical procedure. This problem will be discussed further in the section of software for grey box modelling, see Section 2.4.

From the theory of Linear Quadratic Gaussian systems, we know that the innovation process from a Kalman Filter is a white Gaussian process. Therefore, a Kalman filter can be used to predict the process output when the loss likelihood function is computed from equation (2.8). In addition, the Kalman Filter also computes the covariance matrix of the innovation process.

When a non-linear model is under construction, we have a more complex situation and theory from non-linear filtering has to used to predict the process output. A straightforward way to solve the problem is to use an Extended Kalman Filter. The filter is an approximate computation algorithm used in non-linear systems, (Jazwinsky, 1970). The filter consists of a linearization of the state equations around the estimated state at every discrete sample instant and is an extension of a Kalman Filter for linear systems.

Application of an Extended Kalman Filter to compute the likelihood function is iterated according to the following algorithm, (Camo, 1987):

1. Fetch measured data from a data file, the process output y(k) and process input u(k).

2. Compute the predicted output:

$$\bar{y}(k) = G\left[\bar{x}(k),\ u(k),\ \hat{\theta},\ k\right] \tag{2.9}$$

3. Compute the linearized measurement matrix:

$$\Gamma(k) = \frac{\partial G}{\partial x}\Big|_{x=\bar{x}(k),\ u=\bar{u}(k),\ k} \tag{2.10}$$

4. Compute the covariance of the prediction error:

$$P(k|k-1) = \Phi(k-1)P(k-1|k-1)\Phi(k-1)^T + R_1 \tag{2.11}$$

5. Compute the filter gain:

$$K(k) = P(k|k-1)\Gamma(k)^T\left[\Gamma(k)P(k|k-1)\Gamma(k)^T + R_2\right]^{-1} \tag{2.12}$$

6. Compute the covariance of the estimation error:

$$P(k|k) = \left[I - K(k)\Gamma(k)\right]P(k|k-1)\left[I - K(k)\Gamma(k)\right]^T + K(k)R_2K(k)^T \tag{2.13}$$

7. Compute the estimated states:

$$\hat{x}(k|k) = \hat{x}(k|k-1) + K(k)\left[y(k) - \bar{y}(k)\right] \tag{2.14}$$

8. Compute the predicted states:

$$\bar{x}(k+1|k) = F\left[\hat{x}(k|k),\ u(k),\ \hat{\theta},\ k\right] \tag{2.15}$$

9. Compute the transition matrix:

$$\Phi(k) = \frac{\partial F}{\partial x} \Big|_{x=\hat{x}(k),\ u=u(k),\ k} \qquad (2.16)$$

10. Compute the covariance matrix of the innovation process:

$$\overline{R}(k) = \Gamma(k)P(k|k-1)\Gamma^T + R_2 \qquad (2.17)$$

11. Compute the likelihood function:

$$\mathcal{L}(k|\hat{\theta}) = L(k-1|\theta) + e^T(k)\overline{R}^{-1}(k)e(k) + \ln\left\{\det\left(R(k)\right)\right\} \qquad (2.18)$$

Steps 1 - 11 are repeated from k=1 to k=N. Search procedures to find the parameter vector $\hat{\theta}$ are described in Section 2.4. Note, in the literature there are several variants of the Extended Kalman Filter.

2.3.4 Model analysis

The tentative model has to be analysed after the parameter identification procedure has converged. The grey box model consists of: parts related to the hypotheses of the process, which are formulated during the basic modelling phase, and unsure parts, which are based on the unknown parameters. Furthermore, there are parts of the process which it is not possible to model or as which have a smaller impact on the behaviour of the model. These components are represented as disturbances.

The goal of model analysis is to give a basis to appraise whether the tentative model is good enough and appraise the information how to expand the model. This is not a precise procedure, several weightings have to be taken. It is like starting a trial against the model and investigating what argues in favour and what argues against the model. We also have to estimate the significance of the argumentation. In the following, we give a number of tools to be used for model analysis and we discuss their usefulness.

- **Ballistic simulation.** To find out if the model structure is able to describe the dynamic behaviour of the process. This means that the model is simulated with the same input sequence as used during the experiment. The simulation is made without any predictor. The outputs from the experiment and the simulation are plotted in the same time diagram.

- **Likelihood function tests.** A method for comparing two models is the Likelihood ratio test, (Willks, 1962) and (Rao, 1973). It is a test to investigate if a more complex model is preferred compared to a simpler one. Consider two model sets $\mathcal{M}_1$ and $\mathcal{M}_2$, where $\mathcal{M}_1 \in \mathcal{M}_2$, which means that $\mathcal{M}_2$ is the more complex model. Let the estimates be $\hat{\theta}_1$ and $\hat{\theta}_2$.

When the number of samples is large, we have a measure:

$$\mathcal{L}_R = 2 \cdot N \cdot \left[\mathcal{L}(\hat{\theta}_2) - \mathcal{L}(\hat{\theta}_1)\right] \tag{2.19}$$

which is asymptotically chi-2 distributed, with dim($\hat{\theta}_2$)-dim($\hat{\theta}_1$) degrees of freedom. The threshold for significant change in loss function with a 0.1% risk is shown in table 2.1. There are other structural model tests based on the Likelihood function, discussed by Holst et. alt (1992). For instance the Langrange Multipliers test is a suitable for non-linear models, (Harwey, 1989).

Table 2.1 Threshold for significant loss function with 0.1% risk.

Degree of freedom	*Significant threshold*
1	5.4
2	6.9
3	8.1
4	9.2
5	10.3

- **Parameter range.** When using a physical base for modelling, we have a priori information about the magnitude of the estimated parameters. For example, it is not feasible when an estimate is negative if it should be positive. For instance, a parameter describing the thickness of an emulsion cannot be negative. Often, we also have some idea of the magnitude of the estimates.

- **Parameter identifiability.** We will first define the Hessian as the second derivatives of likelihood function. The Hessian is given as:

$$H(\theta) = \left(\frac{\partial}{\partial\theta}\right)^T \left(\frac{\partial}{\partial\theta}\right) \mathcal{L}(\theta) \tag{2.20}$$

Let us assume that we have found the parameter vector θ_0 for the "true system", i.e. the system which actually generated the data. Fishers information matrix for this case is given as:

$$M = E\{H(\theta_0)\} \tag{2.21}$$

The inverse of Fisher's information matrix gives a lower bound for the covariance of the estimate parameter vector, Cramer-Rao lower bound:

$$\mathrm{Cov}(\hat{\theta}) \geq [M]^{-1}_{\theta=\theta_0} \tag{2.22}$$

The covariance matrix gives insight into the quality of the estimates, and how reliable they are. This has interesting applications in, for example, robust control, where attempts are made to design regulators that are insensitive to parameter variations of a given magnitude, (Graebe, 1990a). The covariance matrix can be used to determine what magnitude of parameter variations one might expect.

Another use is investigation of parameter identifiability. If the covariance of some parameters, found in the diagonal of the matrix, is very large, those parameters are not very sensitive with respect to given measured data. Either the experiment is not sufficiently informative or the model may be over parameterized.

In addition to the diagonal entries of the inverse Hessian matrix, the off-diagonal values indicate coupling among parameters. Large values indicates there is a dependence between estimated parameters.

- **Cross parameter validation.** If possible, the experiment is divided into two sequences, so that we can make the estimation on two different measurement sequences. The circumstances during the experiments should be the same; this means that the treated material is similar and the state of the process is about the same. Based on these presumptions, it should be possible to reproduce the estimates. For each experiment, it is also instructive to plot the estimates on an axis. For example, let the estimated parameter for sequence i be:

$$\hat{\theta}_i = \left\{\hat{\theta}_1\ \hat{\theta}_2\right\} \qquad (2.23)$$

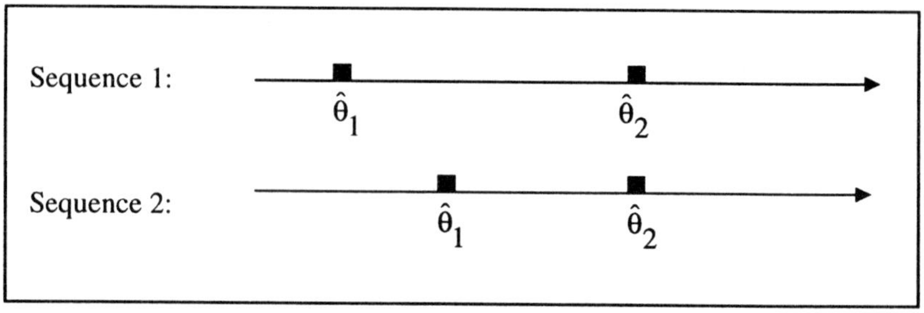

Fig. 2.7 Parameter estimate for two different sequences.

From Fig. 2.7 it is obvious that the estimate of parameter $\hat{\theta}_1$ differs a lot between the two sequences but the parameter $\hat{\theta}_2$ is almost equal. This indicates that we should pay further intention to the part of the model which the first parameter belongs to. The difference in the estimates can originate from wear, unmodelled dynamics or incorrect assumptions about the process.

The parameters can be estimated from a deterministic or stochastic model. When the deterministic version is applied, we fit the unknown parameters "harder" to the actual data sequence, which means that the estimate will differ more between two data sequences than a stochastic model is applied.

- **Sensitive analysis.** It is sometimes useful for identifiability analysis to have a measure of the sensitivity function with respect to the estimated parameters. The most interesting sensitivities are the influence of the change in parameters on the estimated states and predictive outputs. Mathematically, the sensitivities are represented by partial derivatives such as, (Spriet et. alt, 1982):

  $\frac{\partial \hat{x}}{\partial \hat{\theta}}$ Sensitivity of the estimated state with respect to $\hat{\theta}$.

  $\frac{\partial \bar{y}}{\partial \hat{\theta}}$ Sensitivity of the predicted output with respect to $\hat{\theta}$.

  The calculations of the derivatives has to be made numerically. An easy way to do this is by the explicit Euler's method. To find the sensitivity function of the estimated state, we perform a simulation of the process twice. First, the simulation is made with the estimated parameter vector $\hat{\theta}$ and the simulation is repeated with the parameter vector $\hat{\theta} + \Delta\theta$. The sensitivity function is then approximated as:

$$S = \frac{\hat{x}(\hat{\theta} + \Delta\theta) - \hat{x}(\hat{\theta})}{\Delta\theta} \qquad (2.24)$$

  where $\Delta\theta$ is a given alter of the estimated parameter.

  The time plot of the sensitivity function S gives information about the parameter dependence of the model. This information is useful during the parameter identification and while a parameter can be considered as a constant.

- **Residual tests**. The residuals are represented by the innovation process and it is suitable to investigate mismatch between the real process and the process model. The residuals reflect unmodelled parts and disturbances. For the ideal model the innovation process should be white and normally distributed. There should be no correlation between the residuals and the inputs or correlation with any process signals. There are several tests which can be done with the innovation process. The most common ones are: the auto-correlation, the cross-correlation with inputs, (Söderström and Stoica, 1989) and the distribution test which is compared with the normal distribution curve, (Johansson, 1993).

The tests are related to the so-called null hypothesis, which can be explained as the ideal situation concerning the residuals. The null hypothesis is used as a basis for the statistical tests.

The null hypothesis is defined as:

(i) The innovation process is a Gaussian distributed white noise with zero mean.

(ii) The residuals have a symmetrical distribution.

(iii) The residuals are independent of all inputs.

Auto-correlation test.

The auto-correlation of the innovation process is computed as:

$$\hat{r}_{ee}(\tau) = \frac{1}{N-\tau} \sum_{k=1}^{N-\tau} e(k)e(k+\tau) \qquad (\tau \geq 0) \tag{2.25}$$

where τ is the time lag between the residuals. When $\tau=0$, we achieve the estimated value of the covariance of the residuals. The auto-correlation function is often normalised with the covariance and we will have the normalised auto-correlation:

$$\hat{\hat{r}}_{ee}(\tau) = \frac{\hat{r}_{ee}(\tau)}{\hat{r}_{ee}(0)} \tag{2.26}$$

An suitable way to analyse the auto-correlation function is to plot the normalised correlation versus the time lag τ in a diagram. The diagram is complemented with the 99% confidence interval for the asymptotic distribution, which is calculated as $\left[-2.58 / \sqrt{N}\ ,\ 2.58 / \sqrt{N}\right]$.

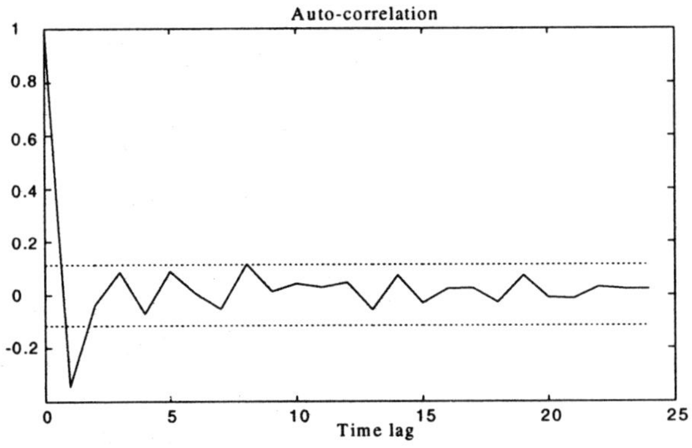

Fig. 2.8 Auto-correlation, *dotted* 99% confidence interval.

An example of the auto-correlation for an innovation process is shown in Fig. 2.8. Since the function has large peaks outside the confidence interval, we can conclude that the residuals are not independent and consist of coloured noise.

Cross-correlation test.

The dependence between the residuals and other parameters is tested in a comparable way to the auto-correlation test. The cross-correlation with the process input is estimated by:

$$\hat{r}_{eu}(\tau) = \frac{1}{N-\tau} \sum_{k=1}^{N-\tau} e(k+\tau)u(k) \qquad (2.27)$$

The cross-correlation function is also normalised:

$$\hat{\bar{r}}_{eu}(\tau) = \frac{\hat{r}_{eu}(\tau)}{\sqrt{\hat{r}_e(0)\,\hat{r}_u(0)}} \qquad (2.28)$$

The cross-correlation test is examined for both positive and negative lag τ. A peak for negative lag indicates that there is a feedback in the system. The process itself can contain a feedback or the experiment is carried out with a control signal which is based on the process output. In the same way as for the auto-correlation test, the cross-correlation plot is compared with the 99% confidence interval.

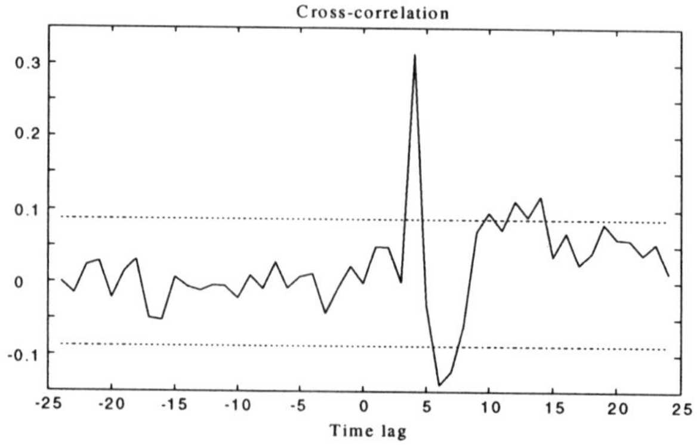

Fig. 2.9 Cross-correlation, *dotted* 99% confidence interval.

In Fig. 2.9, an example of cross-correlation between residuals and input is presented. We can see that there is an peak for positive lags, which may indicated that there is an unmodelled time delay.

<u>*Normal distribution test.*</u>

The correlation tests give information about the independence of the innovation process. Another test is to plot a histogram of the residual magnitudes. The histogram is complemented with a Gaussian curve based on the standard deviation of the innovation process. By comparing the histogram with the normal distribution curve, we have a attractive tool to get information about the residual distribution.

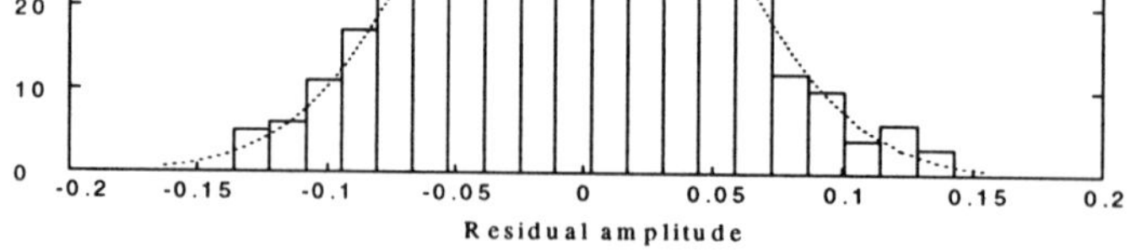

Fig. 2.10 Histogram of residuals, *dotted* the normal distribution.

<u>Residual plot</u>

A simple way to get a quick glance of the residual is just to make time plot and look for trends, peaks or other deviation from what could be called white noise. A "good" residual plot is presented in Fig. 2.11.

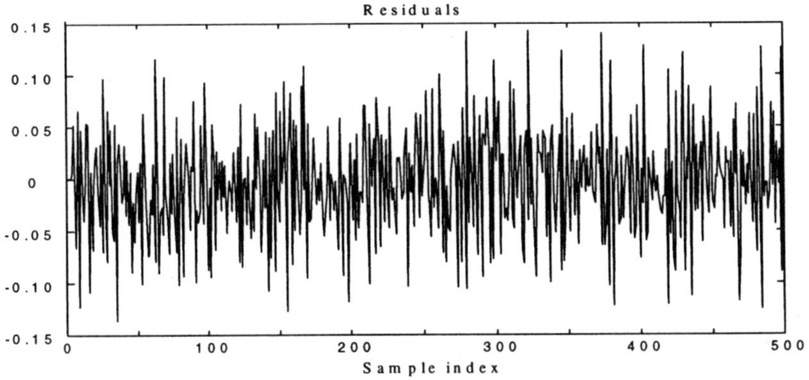

Fig. 2.11 Residual plot.

❐ **Structure tests.** A paramount idea during modelling and identification is to keep the model simple. This means that we need an instrument to compare two different structures ($\mathcal{M}_i$: i=1,2) and a different number of estimated parameters (p_{ei} : i=1,2). Assume there are two model structures $\mathcal{M}_1$ and $\mathcal{M}_2$ such as $\mathcal{M}_1 \not\subset \mathcal{M}_2$ in addition to $p_{e2} > p_{e1}$. The loss function based on $\mathcal{M}_2$ may be lower than that based on $\mathcal{M}_1$, since $\mathcal{M}_2$ contains more parameters. But the question still remains: which structure is most promising? One criterion for handling this problem can be obtained by penalising in some way the decrease of the loss function with the increasing number of estimated parameters; this corresponds to a trade-off between a low value of the loss function and penalty for model complexity.

Akaike (1981) invented the criterion as:

$$\mathrm{AIC} = \mathrm{N} \cdot \log\left[\det \mathrm{R}(\theta)\right] + 2\dim(\theta) \qquad (2.29)$$

2.3.5 Expanded modelling

The basic model described in Section 2.3.1, is based on a priori knowledge about the process. The modelling procedure starts with hypotheses, assumptions and facts concerning the system. This gives a simplified starting model, which may be considered as a core model. However, at least the assumptions about the process may be uncertain but it is also possible that some of the hypotheses can be incorrect. Information about the process are supposed to be correct, but experience of modelling and identification gives that we should also question the information, because the information can arise from tradition and have become facts over a long period of time.

The information may also be wrong due to communication problems, because the model builder collect the information from process operators, process engineer, instructions about the process and so on. During this collections of information there might be misunderstandings, the vocabulary might be different for the people involved or there can be other sources of misunderstanding.

The discussion above indicates that the first attempt to build a basic model might be wrong. Since the basic model is the hub from which we start the further modelling procedure, it is important that we are aware of the problem of incorrect basic models. This means that we must be aware that the basic model is dependent on the a priori information and the simplifications, which may be incorrect

During the model analysis we get information about the behaviour of the model. Further, we must appraise whether the model is good enough to describe the process. The assessment must be done in the light of the purpose of the model. A question arises: should we stop the modelling procedure or should we try to find a better model? Bohlin (1991) discusses the stopping rules when making process models: the scientist's rule and the engineer's rule. The science rule gives a model which describes the physical behaviour as well as possible, but the engineering rule gives a model which describes the process as well the purpose of the model requires. In the industrial situation we are mostly reduced to using the engineering stopping rule.

The expanded modelling phase can seen as expanding the model to a better description of the real process. The expanding procedure should be carried out in small steps so it is possible to analyse the influence of a certain expansion. It might also be possible to expand the model in several directions, so to speak on the same level or on a higher level. The expansion on the same level is made by estimating more parameters or change the influence of some variable. When it is not possible to find a feasible model by this course of action, some new components has to be included in the model which makes the model feasible. Expansion to a higher level is done by enlarge the model with additional parts.

We make a decision based on the model analysis, which direction is the most promising way to proceed. The operation is repeated at several levels until we accept the model. The expanded modelling procedure is illustrated in Fig 2.12.

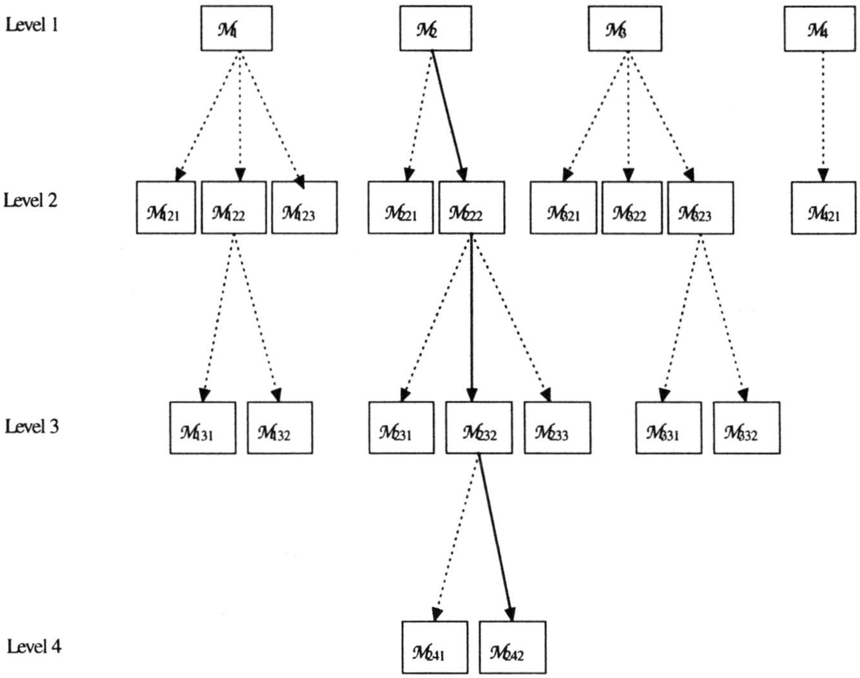

Fig. 2.12 Expanded modelling.

There are four basic models; $\mathcal{M}_1 \ldots \mathcal{M}_4$ presented in Fig. 2.12. Which is the most promising in an engineering sense is difficult to say at this level. The situation can also be that we don't have four different basic models to start with. It is most likely, that we start with the basic model $\mathcal{M}_1$ and expand it further to level 2. For instance, we say that the model $\mathcal{M}_{122}$ is the most promising one and proceed to level 3. Here, we try to expand the model further to the models $\mathcal{M}_{131}$ and $\mathcal{M}_{132}$.

Suppose the achieved models $\mathcal{M}_{131}$ and $\mathcal{M}_{132}$ are not good enough for the purpose, then we have to do something about it. During the model analysis procedure new information about the process is created. We might have made some mistake when the hypotheses, the assumptions or the facts about the process were formulated. From the new information which comes up, we try to formulate alternative basic models. In the example presented in Fig. 2.12 we assume that two more basic models are constructed, models $\mathcal{M}_2$ and $\mathcal{M}_3$. The procedure continues on to level 2. There are now two possible path to take on the modelling procedure. First, start with the basic model $\mathcal{M}_2$ and proceed to level 2, level 3 and level 4. The other way is to start with both basic models $\mathcal{M}_2$ and $\mathcal{M}_3$ and go on to level 2 for both route 2 and 3.

The first method means that route 2 is separately run to the stop, but the second method implies that routes 2 and 3 are investigated in parallel. Which method is to be preferred is an open question. In this example, the basic model $\mathcal{M}_2$ will result in an acceptable model at level 4. But this is not known at the earlier stages, for instance at level 2. Compare this with a crime investigation which proceeds on a broader front in the beginning, i.e. investigates all alternative models at level 1. This will give a more comprehensive picture of the whole situation. This situation demands more of the model builder (or investigator) since more models are considered in parallel. However, more information is generated on each level which can be used at the next level, and maybe strongly lead the model builder to create a "good" model of the process.

In the example given in Fig. 2.12, we accept the model $\mathcal{M}_{242}$. The notation means that the final model originates from basic model $\mathcal{M}_2$, stopping at level 4 and model number 2. The route from the basic model to the final accepted model is represented with a continuous line, while the other routes are dotted.

We shall also discuss how to enlarge a model. It is difficult to give exact guidelines on how to deal with this tricky matter. But some hints can be stipulated. The model analysis procedure gives information which can serve as a basis for a decision on how to expand a model for better performance. It is a matter of how to interpret the analysis information. Fundamentally, the designer is the crucial part of the model building procedure. It is up to him/her to specify alternative model extensions, which can be a scientific or an engineering expansion. The difference lies in the nature of the expansion. The scientific way means the expansion is based on strict scientific knowledge but the engineering way means an explanatory model.

2.3.6 Model appraisal

The expanded modelling procedure gives several models to deal with. The information about the quality of the models is achieved during the model analysis. Since the information is given by several different measures, the question arises how to evaluate the information and how to use it when different models are compared.

During the model extended phase, we can distinguish between two main cases. First, the model can be extended in a nested way; this means that the number of unknown parameters is increased and the model is just expanded. The other case involves a more complicated situation, which means that the model structure is changed and the number of unknown parameters can either be the same or expanded.

In the case of nested model expansion, the likelihood ratio test is favoured, (Bohlin, 1994). For the cases not nested is the problem more difficult. A procedure for this case is suggested by Holst et alt. (1992). They use a Bayesian hypothesis test by introducing a so-called a posterior odds. The decision on which model should be favoured is given by a ratio test from a Bayesian viewpoint. The method is so far not applied to any practical case.

When we have a not nested situation, the problem is to decide which of two or more models should be favoured. For the situation when the number of estimated parameters is the same for the different models, a general rule is to favour the model with the smallest value of the likelihood function.

The most difficult situation to investigate occurs when we are to decide between several models with different structures and different numbers of estimated parameters. When the model structure is expanded to consist of more parameters to estimate, the loss function will decrease. This means it is difficult to decide whether the decrease in the significant loss function comes from a better model structure or expansion of the parameter vector.

Akaike uses the relevance of the information concept during the model building phase and introduced the AIC measure, equation (2.29). The measure tries to give information about which tentative model yields most information concerning the collected data sequences. Other sources for appraisal are given by the model analysis. The ballistic simulation gives valuable information about whether the model can reproduce the dynamics of the process.

In Fig. 2.13 we have placed the model builder in the centre to manage the procedure of model appraisal. He or she is to use all the information gathered during the model analysis and decide in the light of the purpose of the model when it is good enough.

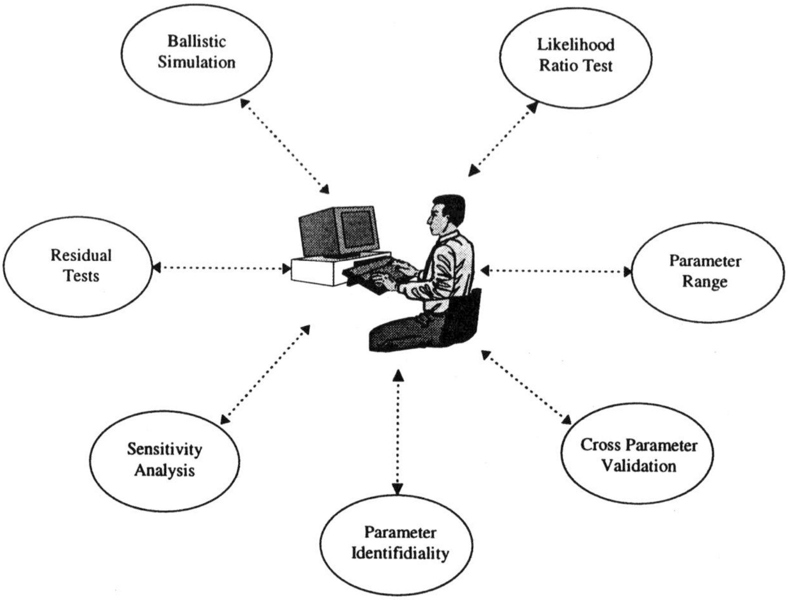

Fig. 2.13 Model appraisal.

2.4 Software tools

When we are dealing with black box identification, we will get an input-output description of the process. A model from a class of standard models (ARX, ARMAX-, Box-Jenkins models and others) is adapted to measured data in a statistical way. The model will typically give a linear relation between the inputs and the outputs. For black box identification there are many program packages available on the market, for example the "System Identification Toolbox" for use with Matlab®.

The process models, which are based on physical insight will often include non-linear relations. This fact demands other properties of the software than is the case with black box identification. Commercially available are, for instance, MATRIX-X® and CYPROS®.

A new software package developed specially for grey box modelling is the IdKit software. The algorithms are described by Bohlin (1984) and the implementation is reported by Graebe (1990a,b,c).

Both the CYPROS and the IdKit software packages use the likelihood method for parameter estimation in a non-linear state space model. The likelihood function is computed by an Extended Kalman Filter. The search routines for the parameters are different and so is also the set of commands handling the procedures

2.4.1 CYPROS software

This software is written in the programming language FORTRAN and consists of a library with routines to handle the identification and the model analysis. This specific software is named KALMAX, (CAMO, 1987). The software package include also routines for:

- Converting ASCII-files to a unique file format used by KALMAX.
- Editing data files. For example handling outliers in the measured data, scaling of data or performing mathematical operations on a data sequence.
- Plotting data on the screen or other plotting devices.
- Statistical operations on a data files. For example plotting auto-correlation and cross-correlation.
- Data filtering. It is possible to perform low pass, high pass and other types of filtering of data sequences.

The Kalmax software is run trough a hierarchy of command levels. The firsts level is called the main command level. In the following a brief description of each main command is given.

- **Initiation.** Before one can enter the main level, an initial mode must be passed, which implies, the number of states, inputs, outputs and unknown parameters is specified. We must also give the name of the data file and the name of data base file which is used by the software to save other information needed for identification.

- **Optimisation.** Before optimisation can start, we must define which parameters are to be estimated and the parameter error bound used by minimisation routine. The Likelihood function is then minimised based on Powel's method, (Powel 1978). The parameter search can be followed on the screen and the optimisation can be interrupted whenever desired.

- **Simulation.** There are three commands below this main level*; simulation, ballistic simulation and sensitive analysis.* The simulation is made with an Extended Kalman Filter. The ballistic simulation is made without updating the states, i. e. the Extended Kalman Filter gain is set to zero. The sensitive analysis is made by simulation of the system twice with the Extended Kalman Filter and with a perturbation of the parameter estimates.

- **Curvature analysis.** This contains calculation of an approximation of the Hessian, its eigenvalues and eigenvectors. The inverted Hessian is also calculated.

- **Defining.** Under this main command three other commands are available. These are *storage, filter type and criterion type*. By storage we can define which parameters are to be stored during simulation. From the filter type command we can choose between using a complete Extended Kalman Filter, a partial constant Extended Kalman Filter which uses a constant filter gain after a specified number of samples, a user-defined filter which makes it possible for the model builder to specify his own filter for updating the states.

- **Matrix manipulation.** There are three commands beneath this main level; *change, store and fetch.* The command *change* clearly changes the information used for identification. The *store* command writes the information to the data base and *fetch* reads the information from the data base.

2.4.2 MoCaVa / IdKit software

The IdKit package is a result of an ongoing research project at the Royal Institute of Technology, department of Signals, Sensors and Systems. The concept was originally initiated by Bohlin (1970) and expanded with comprehensive algorithm specification in Bohlin (1984). A software implementation of these algorithms was performed by Graebe (1990a, b, c). The software is written in the programming language C.

MoCaVa (Model Calibration and Validation) is a user-shell developed for IdKit. It breaks down the task of grey box modelling into three parts, invoked by the commands:

- **Predat.** That is a routine for pre-formatting a data batch. It gives the user an opportunity to view statistics of the data batch. Some initiation is also done such as scaling and naming.

- **Calibration**. Identification of a process is performed in this interactive routine. The model of a process is achieved by recursively fitting and testing a series of tentative model structures. Starting from the "simplest" model the structure of the model is expanded by adding submodels that significantly contribute to the reduction of the overall loss. The method of "pruning and cultivation" to expand a model structure is discussed in (Bohlin 1994). There Bohlin derives a designer's guide for grey box modelling which is now implemented as the Calibration routine.

- **Validation**. Validation should always be interpreted in the sense of how the model will be used. Therefore very few papers concerning validation have been written, leaving the subject of determining the models value to the user. In spite of this, MoCaVa supports the Validation routine, where the model can be validated in terms of reproducibility, (short- or long range) prediction, prediction of measured and unmeasured variables and parameter accuracy. Note that it is still up to the user to determine whether or not these measures are adequate for the purpose of the model.

The identification is based on likelihood function in a similar way as the Kalmax package. However, the search routine is quite different. It is based on the Newton-Raphson method and the parameter vector is updated by using the inverted Hessian and the derivatives of the Likelihood function. The routine is formulated as:

$$\hat{\theta}(k+1) = \hat{\theta}(k) - \left\{ H(\theta)^{-1} \left[\frac{\partial \mathcal{L}(\theta)}{\partial \theta} \right] \right\} \Bigg|_{\theta = \hat{\theta}(k)} \qquad (2.30)$$

❐ **Remarks on the MoCaVa / IdKit software**

- The process can be modelled as a system of non-linear, deterministic state-vector model with stochastic or deterministic inputs. It accepts noisy and sparse data.
- A stiff ODE solver is implemented, giving the user the possibility to solve stiff problems. A drawback in IdKit is that it does not support ADE, however these algebraic loops can often be formulated as fast dynamic states and hence exploit the stiff ODE solver.
- MoCaVa separates explicitly the model of the system that generated the data into the three blocks: sensor, process proper and environment, in the double interest of facilitating the modelling of random disturbances (if any), and increasing computing efficiency.
- The package is "open" in the sense that it accepts user submodel definitions, or library models from an external source able to produce C-code. This facilitates reuse of submodels.
- As a platform, a UNIX operating system with a C-compiler and the "curses" and "X" libraries for user interface is used. This provides portability. It also simplifies the assembling of submodels into a system, since the C-linker provides the services needed for the task.
- Several plotting facilities provide a good support for understanding the process in the different routines.

2.5 Summary

In this section, we gave a presentation of the grey box modelling method. It is formulated as several different activities: basic modelling, experiment, expanded modelling, identification, model analysis and model appraisal. The model builder plays an active role from defining the purpose of the model to forming a useful model.

Basic modelling includes several steps which guide the model builder from a general description to a basic discrete time model. The general description means a document which describes the function of the process. It includes physical behaviour of the process and all the information which is possible to acquire from different sources. The result of the basic modelling procedure is a set of equations which are possible to code to a software module.

The *experiment* is one of the most important activities, since it gives the basis for the further actions. New information about the process can originate from the measured data and is a prerequisite for the identification and model analysis.

The *identification* procedure is based on maximising the likelihood function. This method is suitable for identification of non-linear models. The two software tools presented in this book are based on the Extended Kalman Filter.

Model analysis consists of several tools, which give information about the model. In this chapter, we discuss the following: ballistic simulation, likelihood function test, parameter range, parameter identifiability, cross parameter validation, sensitive analysis, residual tests and structure test.

The *expanded modelling* procedure is used to refine the basic model by incorporating new knowledge, which is generated from the other activities. The expanded modelling phase can seen as expanding the model to a better description of the real process. The expanding procedure should be carried out in small steps so it is possible to analyse the influence of a certain expansion.

The last procedure is the *model appraisal*, which means a comparison of the results from the model analysis. During this activity, we have to take into account the purpose of the model, since the purpose decides the necessary characteristics of the model.

3 APPLICATION OF BASIC MODELLING

3.1 Introduction

The production of steel strip at the SSAB Steel Plant consists of several different processes, from heating slabs to hot rolling, pickling, cold rolling, and so on, depending on what kind of product is ordered. The material flow for the production line from heating to the finishing mill is shown in Fig. 3.1.

The slabs are reheated in a continuous reheating furnace to a temperature of about 1100 °C. A hot rolling mill with six stands reduces the thickness of the slabs to between 2 and 10 mm. During hot rolling, oxide scale grows on the iron surface. The thickness of the scale depends on the circumstances for the hot rolling, for example the cooling condition. After hot rolling, the steel strip is reeled in coils. The coils are cooled for about two days in a storage area. Afterwards, the steel strip proceeds to the pickling line. The steel temperature at the inlet to the pickling line varies depending on the size of the coil and the outdoor temperature. This also influence the amount of oxide scale on the steel strip surface.

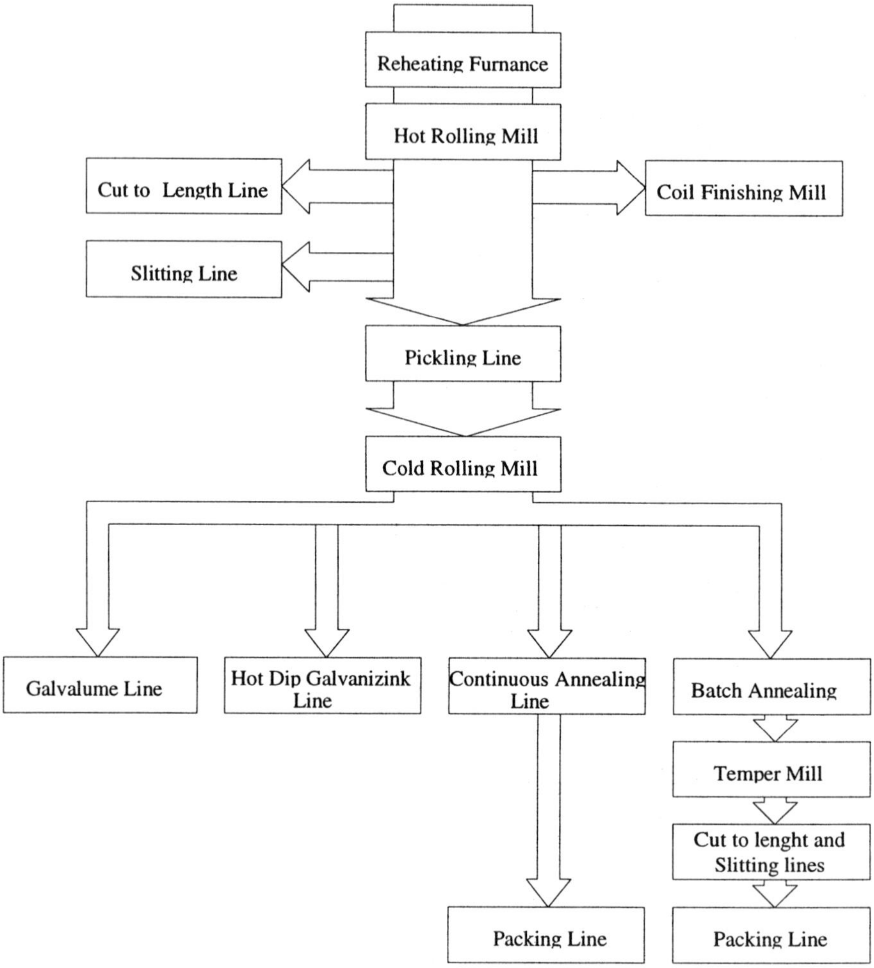

Fig. 3.1 Material flow at SSAB Steel Plant.

After passing the pickling line the steel strip can be treated in a galvalume line, a hot dip galvanizink line, a continuous annealing line or a batch annealing line. In this book we focus on the pickling line and specially the rinsing process within the line.

The pickling line consists of a pickling process, which is divided into four separate tanks of hydrochloric acid, and a rinsing process, which is divided into five separate rinse tanks. The steel strip passes continuously, first the pickling process and directly afterwards the rinsing process. In the pickling process, scale and other impurities are removed from the surface of the steel strip. After pickling, the steel strip is rinsed by passing through the rinsing process.

In Fig. 3.2 we can see that the rinse tanks are physically situated directly after the acid tanks. A part of the hydrochloric acid is transferred via the steel strip from the tanks of hydrochloric acid to the rinse tanks and mixed with the rinse liquid. As a result of the rinse process a waste product of polluted water is produced, which must be treated and disposed afterwards.

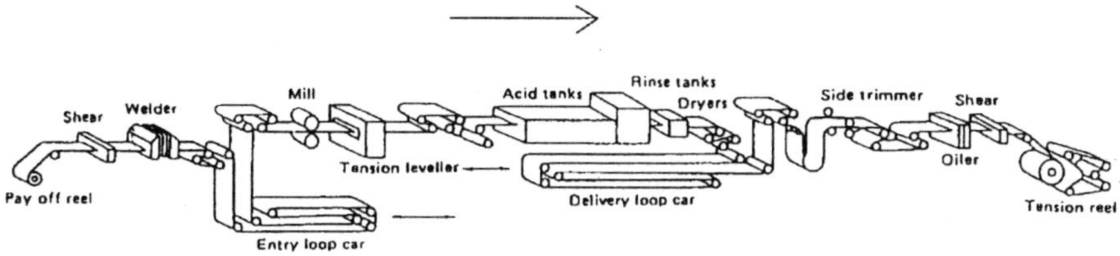

Fig. 3.2 The pickling line.

In the entry part of the pickling line, the ends of the steel strip are welded together to form a continuous steel strip. The feeder stage consists of an entry loop car, see Fig. 3.2, which makes it possible to weld the ends together continuously. The exit stage also consists of a delivery loop car, which makes it possible to divide the steel strip into coils again without stopping the process.

After leaving the pickling process the steel strip passes between a double pair of squeezer rolls. The purpose of the squeezer rolls is to reduce the flow of liquid carried with the steel strip. The steel strip is not completely dried, but retains a thin film of liquid. This film consists of the same chemical liquid as in the last pickling tank, mainly hydrochloric acid and a small portion of dissolved iron.

If hydrochloric acid were left on the steel strip, the hydrochloric acid could react with the iron. When the water and the hydrochloric acid on the steel strip vaporise, the rest of the film consists of more and more iron, and iron oxide is formed on the surface of the steel strip. Visually, the steel strip becomes discoloured with a brown emulsion. Furthermore, the attack on the steel strip will take place in the presence of chloric ions. In contact with air, the steel strip will rust heavily. Consequently, the steel strip must be rinsed well after pickling to prevent precipitation of iron hydroxide and to prevent the steel strip from rusting.

In this chapter, we deal with application of basic modelling of a steel strip rinsing process. The procedure described in Chapter 2 is used and we penetrate the procedure in detail.

3.2 Basic modelling of the rinsing process

3.2.1 Process description

The rinsing process is divided into five separate zones, see Fig. 3.3. The acid is transferred via the strip from the last acid tank into rinse zone 1. In each zone the strip is rinsed by a circulating flow of rinse liquid. The rinse liquid is taken from the bottom of the rinse tank and sprayed at high pressure on the top and bottom sides of the steel strip. The rinse liquid falls in drops back into the rinse tank. This means that the film liquid, which is left on the steel strip has a lower concentration of hydrochloric acid at the end of the rinse zone than at the beginning of the rinse zone. The procedure is repeated when the steel strip passes the other rinse zones, and the steel strip becomes cleaner after every rinse zone. Consequently, the film liquid contains a decreasing amount of hydrochloric acid as the steel strip passes through the rinsing process.

After every rinse zone, the steel strip passes between a pair of squeezer rolls. The rolls reduce the amount of rinse liquid transferred via the steel strip to the next rinse zone. At the entry to rinse zone 1 and at the exit of rinse zone 5, the steel strip passes between double pairs of squeezer rolls. The principle construction of the rinsing process is shown in Fig. 3.3.

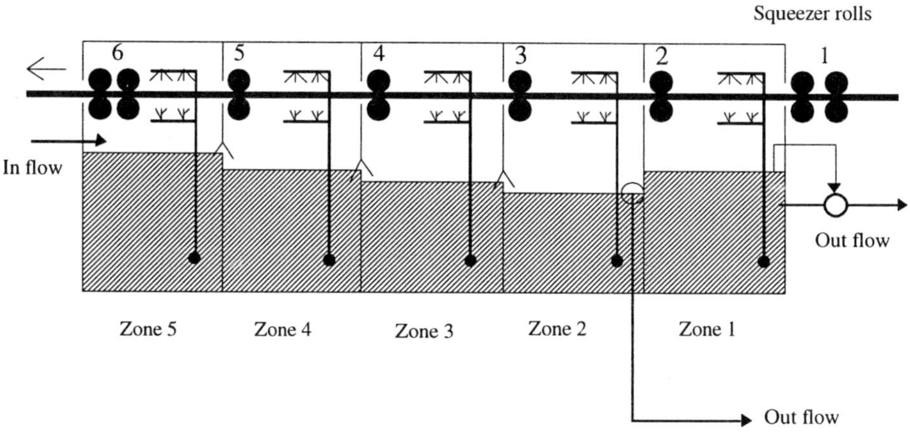

Fig. 3.3 The rinsing process.

The squeezer rolls are made of iron with a surface of rubber. When the process is running, the surface of the rolls wears. This means that the flow via the steel strip will increase and the concentration in the following tanks will increase. If the squeezer rolls become worn out, the risk of a badly rinsed strip is apparent.

During planned maintenance stops, worn rolls are replaced by renovated rolls. These stops normally happen every second week. Apart from normal wear of the squeezer rolls, large parts of the rubber can come loose, although this happens more seldom. As this is a serious matter, the production is stopped and the damaged squeezer roll is replaced.

Depending on several production variables such as strip velocity, strip width and strip thickness, the amount of hydrochloric acid transferred to the following tanks varies. This means that the concentrations in the tanks vary depending on the production variables. Consequently, the cleanness of the strip after passing rinse zone 5 will vary, because the cleanness of the strip depends on the acidity of the rinse liquid.

Clean water is fed into tank 5. This is done to dilute the rinse liquid in this tank, because otherwise the concentration of hydrochloric acid would become too high. An equivalent amount of rinse liquid flows from tank 5 to tank 4. This means that the rinse liquid in tank 4 is mixed with rinse liquid from tank 5 with a lower concentration. The flow of rinse liquid proceeds in the same way to the following tanks through tank 3 and tank 2.

The main part of the rinse liquid is drained off from tank 2. Only a small part is pumped from tank 2 to tank 1. For tank 1 it can be pointed out that the flow from this tank is constant, 0.3 [m^3/h], while the flow into tank 1 is controlled so the level of liquid is kept constant. This control system is based on a PI-controller. The variation of liquid level has been measured to +/- 0.05 [m] around the reference level.

The different rinse tanks are separated from each other by walls. In the wall a section is removed, so the rinse liquid can flow to the next tank. The section consists of three oval holes. The construction of the wall, seen from inside the tank is shown in Fig. 3.4.

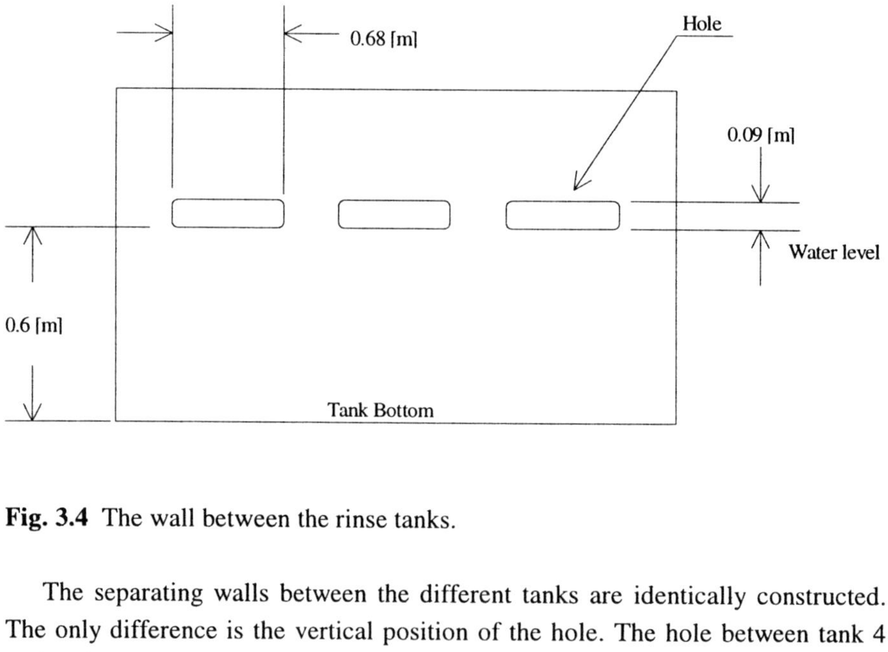

Fig. 3.4 The wall between the rinse tanks.

The separating walls between the different tanks are identically constructed. The only difference is the vertical position of the hole. The hole between tank 4 and tank 3 is at a lower position than the slit between tank 5 and tank 4, see Fig. 3.3. This means that the surface of the rinse liquid is at a lower level in tank 4 than in tank 5. The position of the hole is located so the flow of rinse liquid can only go in one direction through the holes. In Fig. 3.3 it can be pointed out that the level of rinse liquid in tank 1 is higher than in tank 2. This is because the wall between these tanks does not have any openings. The rinse liquid is pumped from tank 2 to tank 1.

The next tank always has a lower level of rinse liquid than the preceding tank. Consequently, the rinse liquid flows freely into the next rinse tank. When there is no flow of rinse liquid through the rinsing process, the lower edge of the hole decides the level of rinse liquid in the each rinse tank.

As described earlier in this chapter, rinse liquid is sprayed under high pressure at the steel strip. This means that rinse liquid splashes heavily inside the rinse zone. The risk is great that the rinse liquid will splash through the slits in the walls. The slits, are therefore, shielded by a cover, which prevents rinse liquid from splashing through the slit. In Fig. 3.5, the slit cover is shown as an upside down V. The slit cover encloses the slit at an angle of 30 degrees.

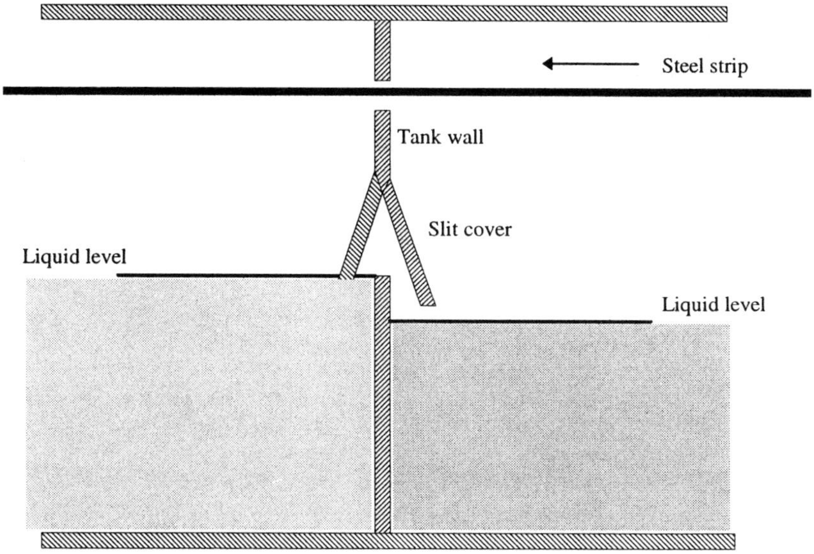

Fig. 3.5 Slit with slit cover.

The flow of clean water into tank 5 is controlled by a separate controller. The aim of this control system is to maintain a concentration in tank 5 below a specified limit. This limit is chosen so that there will be no problem with precipitation of iron hydroxide or with rust on the steel strip. In extreme situations there is a possibility to feed an extra supply of clean water into tank 2.

When the process is running, the strip velocity varies frequently. These variations depend on different production situations. There can be problems when connecting the ends of the steel strip due to the welding taking a longer time than is possible to compensate for, using the storage in the entry loop car. This means that the strip velocity has to be decreased during the welding phase. For the same reasons there are sometimes short stops of the steel strip. During these stops the circulated rinse flows are shut off, because the steel strip will not become any cleaner even though the strip is stopped. The flow of clean water into tank 5 is also shut off during a stop. It is not considered necessary to feed clean water into the rinsing process when there is no flow of hydrochloric acid into the rinsing process.

It is debatable whether it would not be better to feed clean water into the process even during the shorter stops. If the concentration in the tanks is too high, the concentration should be decreased as soon as possible, even if the steel strip is stopped.

The whole pickling process including the rinsing process is built as an enclosed system. To prevent evaporation of hydrochloric acid out into the factory building, there is an air sweating system installed in the pickling line. This system takes care of evaporation of hydrochloric acid from the whole pickling line. This is done by air being drawn from the pickling line and being built up an under pressure inside the pickling line. The moisture is taken care of by a separate process. The water vapour is condensed and the air is cleaned before it is let out. Air is admitted into the pickling line mainly at the beginning and at the end of the process. The system is not hermetically closed; in fact air seeps in through narrow openings along the whole process.

3.2.2 Process structure

Every rinse zone is regarded as a separate subsystem within the rinsing process. The subsystems are connected by the different flows appearing within the rinsing process. The relation between the flows of liquid is shown in Fig. 3.6, (Stein, 1988).

The flows as shown in the figure consist of four different types. F_{m1} to F_{m5} represent the main flow of rinse liquid between the tanks, F_{c1} and F_{c2} are the controlled flow of clean water fed into tank 5 and tank 2. F_{b1} to F_{b6} are the flows via the steel strip through the process and F_{a1} to F_{a5} are the flows of evaporation from each rinse tank. The flow via the strip seems to be affected by several production variables as: strip velocity and strip width. The acid concentration is measured in each rinse tank by conductivity transducers.

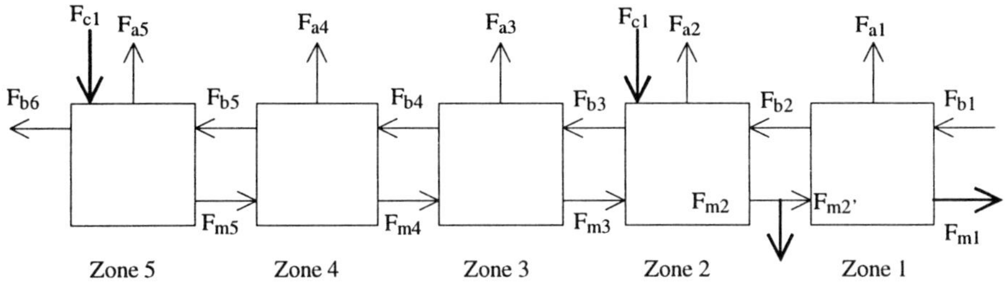

Fig. 3.6 Flows within the process.

3.2.3 Process hypotheses

To be able to form a basic model, we have to formulate a number of hypotheses for the process. Otherwise, the basic model will be too complex and it will be difficult to find a starting point during the modelling work. We must also keep in mind that a hypothesis should be possible to verify, by pre-calculation or by identification with measured data. Note the distinction between hypotheses, assumptions and facts. Assumptions are not verified, and facts are always corrects.

❒ **Hypothesis $\mathcal{H}_1$.**

The first hypothesis concerns the type of process. We consider the rinsing process to be a mixing process with different flows, which contain hydrochloric acid in various concentrations. This means that no chemical reaction takes place.

❒ **Hypothesis $\mathcal{H}_2$.**

The mixing in the tanks is perfect. This means that the concentration in the rinse tank is uniform and independent of the depth or length of the rinse tank. The reason for perfect mixing is that the rinse liquid is taken from the bottom of the rinse tank and sprayed on the steel strip within the rinse zone. This circulation rinsing flow is about 100 [m^3/h]. The volume of the rinse tank is between 5 -10 [m^3], depending on which tank is considered. Consequently, it takes about 3 - 6 minutes to turn over the whole volume in a rinse tank. From this point of view, we can consider the rinse liquid to be completely stirred.

❒ **Hypothesis $\mathcal{H}_3$.**

The worst rinsing situation happens when the steel strip passes the rinse process with the highest velocity. In such cases, we have to discuss whether the strip is completely rinsed after passing a rinse zone or if there will be a retained concentration from the previous rinse tanks on the steel strip. The maximal velocity of the strip is 4 [m/s]. The length of the rinse zone is about 4 meters. During this time, a meter of the strip is sprayed with 28 litre of rinse liquid. This corresponds to three or four buckets of rinse liquid, which we can consider as being rather a lot. The hypothesis means that the concentration of the film liquid on the strip when it leaves a rinse zone has the same concentration as the rinse liquid in the tank.

❒ **Hypothesis $\mathcal{H}_4$.**

Because the rinse liquid is hot, the rinse liquid evaporates from the rinse zones. The rinse liquid in the tanks mainly consists of a mixture of water and hydrochloric acid. The composition of the vapours has the same proportions as the rinse liquid. The amount of evaporated liquid is computed from an empirical formula (Instruction Pickling line - SSAB):

$$F_a = (19.2 + 13.6Va)(Z_{ab} - Z_{af})A_a \qquad (3.1)$$

F_a	Evaporation	[kg/h]
V_a	Velocity of the air above the bath surface	[m/s]
A_a	Bath area	[m^2]
Z_{ab}	Liquid in the air above the bath in the zone	[Kg liquid/Kg air]
Z_{af}	Liquid in the air in the factory	[Kg liquid/Kg air]

For the air in the factory, a temperature of twenty degrees Celsius and a relative humidity of 60% is assumed. From a Mollier diagram, the value of Z_{af} is estimated to $Z_{af}=0.01$. The air temperature in the rinse zones is approximately sixty degrees Celsius. A Mollier diagram gives $Z_{ab}=0.1$.

With a bath area of 9.0 [m^2] and an air velocity above the bath area of 0.1 [m/s], equation (3.1) gives an evaporation of approximately 0.017 [m^3/h]. The value of the air velocity is estimated from the fact that the stream of air into the process through existing narrow openings must at least be 1.0 [m/s]. A total section of all these narrow openings is much smaller than a cross section of the air stream above the bath area within the rinse zone. The evaporation from the spray inside the rinse zone is much more difficult to estimate. A reasonable assumption of the evaporation from the spray is that it is equal in amount to the evaporation from the bath area. It must also be pointed out that the process is built in an enclosed system with roof and walls. These areas are colder than the humid air within the rinse zone. This means that evaporated liquid is condensed at the top and walls of the rinse tank and afterwards falls in drops back to the rinse tank.

The total amount of evaporated liquid from each rinse tank is the sum of the evaporation from the bath and the spray. This means that the total evaporation is less than 0.04 [m^3/h]. This agrees with the estimation made by Kushner (1976). We have made an attempt to estimate the evaporation from measured data. This estimation gave small values of evaporation. The main flow of rinse liquid between the rinse tanks is greater than 1.0 [m^3/h]. This means that the flow of evaporation is much smaller than the main flow. Consequently, we neglect the influence of the evaporation in the model of the process. It has also been shown by Bohlin (1991b) that the model will not improve if evaporation is included.

❐ **Hypothesis $\mathcal{H}_5$.**

Here the dynamic of the main flow is analysed. Clean water is fed into tank 5. As a consequence, there will be a flow of rinse liquid from tank 5 to tank 4; further to tank 3 and tank 2. Most of the rinse liquid is drained off the process after tank 2; just a minor part is pumped over to tank 1. The flow dynamic of tank 1 is consequently of less interest.

To study the dynamics of the main flow, we use the principle of mass balance. As discussed in the previous section, every rinse tank has a basic volume, which is decided by the lower edge of the orifices in the walls between the tanks. Above this volume, there is a small increase, originating from the main flow between the tanks. These increases in volumes in tank 2 to tank 5 are notified by ΔV_2 to ΔV_5.

From the discussion in hypothesis $\mathcal{H}_4$, the flow from evaporation is small compared to the main flow; consequently we let the evaporation be equal to zero. Furthermore, we consider here the difference between inflow and outflow via the steel strip to be small compared to the main flow and to have a small influence on the dynamics of the main flow.

Application of mass balance for the liquid in each tank gives, see Fig. 3.6 for notations:

$$\frac{d\Delta V_5}{dt} = F_{c1} - F_{m5} \qquad \text{(Tank 5)} \qquad (3.2)$$

$$\frac{d\Delta V_4}{dt} = F_{m5} - F_{m4} \qquad \text{(Tank 4)} \qquad (3.3)$$

$$\frac{d\Delta V_3}{dt} = F_{m4} - F_{m3} \qquad \text{(Tank 3)} \qquad (3.4)$$

$$\frac{d\Delta V_2}{dt} = F_{m3} - F_{m2} \qquad \text{(Tank 2)} \qquad (3.5)$$

For the increase in volume which causes the main flow between the tanks, we have that the increase in volume is equal to the increase in the liquid level times tank area:

$$\Delta V_i = \Delta h_i \cdot A_i \qquad (i=2\ldots5) \qquad (3.6)$$

The main flow between the tanks can be considered as a free stream of water into air and as a free fall through a rectangular opening. From Handbook of Engineering, the main flow is computed by, (b is the total width of the orifices in the walls):

$$F_{mi} = \frac{2}{3}\lambda \cdot b \cdot \Delta h_i \sqrt{2 \cdot g \cdot \Delta h_i} \qquad (i=3\ldots5) \qquad (3.7)$$

The flow from tank 2 consists of two parts; one part is pumped over to tank 1 and one part flows through a circular hole in the wall of tank 2. This opening has a radius R_c. The flow can be seen as free fall and the opening will never be completely full when the liquid flows through the hole. The constant λ depends on the form of the opening and g is the constant of gravitation. The hole is shown in Fig. 3.7. Free fall through a circular opening is computed by, (Handbook of Engineering):

$$F_{m2} = \lambda\sqrt{2 \cdot g} \int_0^{\Delta h_2} y\sqrt{h}\, dh \qquad (3.8)$$

Consider an area of height Δh and width y. The area is placed at a vertical level of h related to the lower edge of the opening, see Fig. 3.7.

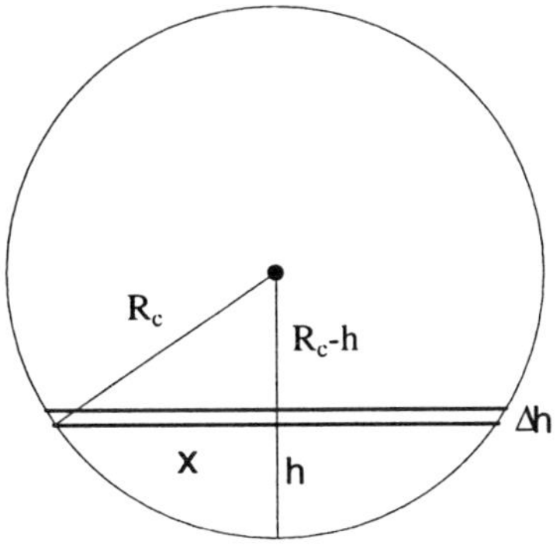

Fig. 3.7 Flow through a circular area.

We have from Fig. 3.7:

$$y=2x \qquad (3.9)$$

$$x^2 + (R_c-h)^2=R_c^2 \qquad (3.10)$$

Because $h << R_c$ we have from equation (3.10):

$$x = \sqrt{2 \cdot h \cdot R_c} \qquad (3.11)$$

Equations (3.8), (3.9) and (3.11) give:

$$F_{m2} = \lambda \cdot 2\sqrt{R_c \cdot g}\,\Delta h_2^{\frac{2}{2}} \qquad (3.12)$$

To sum up, the equations which model the main flow through the rinsing process are described by:

$$\frac{d\Delta h_5}{dt} = \left[c_{hs} \cdot F_{c1} - a \cdot \Delta h_5^{3/2}\right] / A_5 \qquad (3.13)$$

$$\frac{d\Delta h_4}{dt} = a \cdot \left[\Delta h_5^{3/2} - \Delta h_4^{3/2}\right] / A_4 \qquad (3.14)$$

$$\frac{d\Delta h_3}{dt} = a \cdot \left[\Delta h_4^{3/2} - \Delta h_3^{3/2}\right] / A_3 \qquad (3.15)$$

$$\frac{d\Delta h_2}{dt} = \left[a \cdot \Delta h_3^{3/2} - e \cdot \Delta h_2^{3/2} + c_{hs} \cdot F_{c2} - c_{hs} \cdot F_{m2}'\right] / A_2 \qquad (3.16)$$

where:

c_{hs}=1/3600 (Time conversion from hours to seconds).

$$a = \frac{2 \cdot \lambda \cdot b\sqrt{2 \cdot g}}{3}$$

$$e = \lambda \cdot 2\sqrt{R_c \cdot g}$$

We know by experience that the time constant of the mixing process is long. The time constant of the mixing process is approximately the volume of the tank divided by the main flow. If the volume of the tank is 6.0 [m^3/h] and the main flow is 3.0 [m^3/h] the time constant is 2.0 [h].

To investigate whether it is possible to consider the main flow as a stiff system compared to the mixing process, we simulate the flow between the tanks. During the simulation, we let the flow of clean water fed into tank 5 change as a step from F_{c1}=1.0 [m^3/h] to F_{c1}=3.0 [m^3/h]. Furthermore, we assume that no rinse liquid is pumped from tank 2 to tank 1 and F_{c2}=0.

This is not a serious restriction. The simulation starts from the steady state. Normally, the flow friction λ is between 0.8 and 1.0. Without any flow friction we have that $\lambda=1.0$. In the practical situation there is a flow friction, which means that $\lambda<1.0$. A reasonable assumption is $\lambda=0.8$.

Comments on the simulation.

In Fig. 3.8, the result of the simulation is shown. The flow from tank 5 reaches the final value faster than the flow from the other tanks; this is natural because the tanks are connected in a cascade. The flow from tank 2 has a different shape, since the drain opening has a different form than the orifices on the other walls.

In a comparison between the dynamic of the mixing process and dynamic of the main flow, it is seen from Fig. 3.8 that the rise time of the main flow is much shorter than that of the mixing process. Consequently, the main flow can be considered as a stiff system compared to the mixing process.

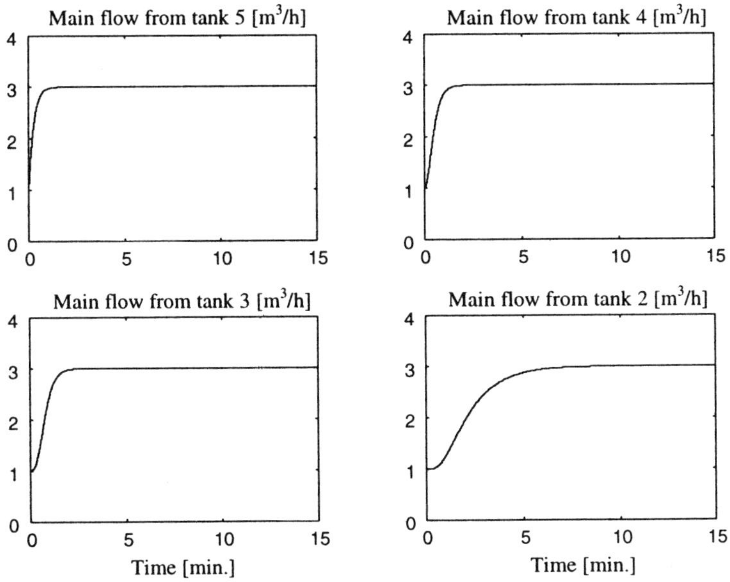

Fig. 3.8 Main flow between rinse tanks.

3.2.4 Basic equations

To model the concentration of hydrochloric acid in the rinse tanks, we apply the principle of mass balance for each rinse tank, (Kushner, 1976). From the discussion about hypothesis $\mathcal{H}_5$, the main flow can be considered as a stiff system. Furthermore, according to hypothesis $\mathcal{H}_4$ we neglect the influence of evaporation from the rinse zones and the flow via the steel strip from zone 5.

The flows which are important for the process are shown in Fig. 3.9 and the concentrations in rinse tank 1 to rinse tank 5 are represented by C_1 to C_5.

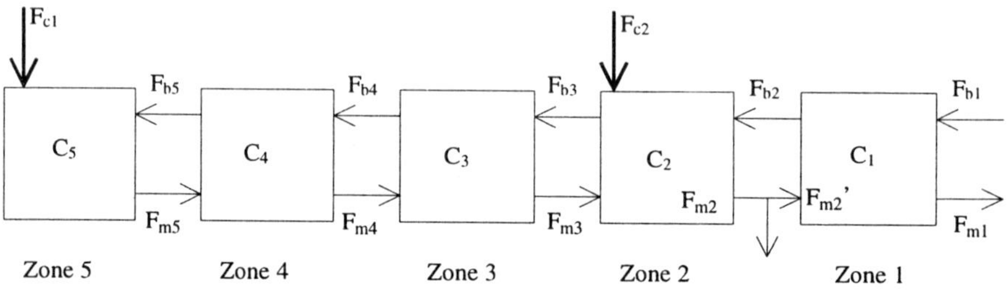

Fig. 3.9 Flows and concentrations within the process.

The mixing in the rinse tanks is assumed to be complete, see hypothesis $\mathcal{H}_2$. This means that the concentration is the same in the tank independently of which part is considered. From hypothesis $\mathcal{H}_3$, the steel strip is also assumed to be completely rinsed after the strip leaves a rinse zone. This means that the film liquid on the steel strip at the end of the rinse zone has the same concentration as the rinse liquid in the tank.

According to the principle of mass balance based on hypothesis $\mathcal{H}_1$, we obtain the following equations to describe the mixing in tank 1 to tank 5:

$$\frac{d(V_1 \cdot C_1)}{dt} = F_{b1} \cdot C_0 - F_{b2} \cdot C_1 + F_{m2}' \cdot C_2 - F_{m1} \cdot C_1 \tag{3.17}$$

$$\frac{d(V_2 \cdot C_2)}{dt} = F_{b2} \cdot C_1 - F_{b3} \cdot C_2 + F_{m3} \cdot C_3 - F_{m2} \cdot C_2 \tag{3.18}$$

$$\frac{d(V_3 \cdot C_3)}{dt} = F_{b3} \cdot C_2 - F_{b4} \cdot C_3 + F_{m4} \cdot C_4 - F_{m3} \cdot C_3 \tag{3.19}$$

$$\frac{d(V_4 \cdot C_4)}{dt} = F_{b4} \cdot C_3 - F_{b5} \cdot C_4 + F_{m5} \cdot C_5 - F_{m4} \cdot C_4 \tag{3.20}$$

$$\frac{d(V_5 \cdot C_5)}{dt} = F_{b5} \cdot C_4 - F_{m5} \cdot C_4 \tag{3.21}$$

The change in volume belonging to the main flow is very small compared to the total volume of the tanks. Consequently, it is possible to consider the volumes as constant.

$$\frac{dC_1}{dt} = \left(F_{b1} \cdot C_0 - F_{b2} \cdot C_1 + F_{m2}' \cdot C_2 - F_{m1} \cdot C_1\right) / V_1 \tag{3.22}$$

$$\frac{dC_2}{dt} = \left(F_{b2} \cdot C_1 - F_{b3} \cdot C_2 + F_{m3} \cdot C_3 - F_{m2} \cdot C_2\right) / V_2 \tag{3.23}$$

$$\frac{dC_3}{dt} = \left(F_{b3} \cdot C_2 - F_{b4} \cdot C_3 + F_{m4} \cdot C_4 - F_{m3} \cdot C_3\right) / V_3 \tag{3.24}$$

$$\frac{dC_4}{dt} = \left(F_{b4} \cdot C_3 - F_{b5} \cdot C_4 + F_{m5} \cdot C_5 - F_{m4} \cdot C_4\right) / V_4 \tag{3.25}$$

$$\frac{dC_5}{dt} = \left(F_{b5} \cdot C_4 - F_{m5} \cdot C_4\right) / V_5 \tag{3.26}$$

For the flows to each tank respectively we apply the principle of mass balance and based on hypothesis $\mathcal{H}_5$ the main flow is considered as a stiff system. Consequently, we get five algebraic equations for tank 1 to tank 5.

$$F_{b1} - F_{b2} + F_{m2} - F_{m1} = 0 \tag{3.27}$$

$$F_{b2} - F_{b3} + F_{m3} - F_{m2} + F_{c2} = 0 \tag{3.28}$$

$$F_{b3} - F_{b4} + F_{m4} - F_{m3} = 0 \tag{3.29}$$

$$F_{b4} - F_{b5} + F_{m5} - F_{m4} = 0 \tag{3.30}$$

$$F_{b5} + F_{c1} - F_{m5} = 0 \tag{3.31}$$

In the basic equation there are parts which are described as separate variables and have no other connections. We have to make assumptions about how to relate them to other parameters. For the flow via the strip, we assume that most is carried by the top and bottom side of the strip and the film thickness has a certain thickness. The flow via the strip can then be expressed as:

$$F_{b1} = K_b \cdot B_b \cdot B_v \tag{3.32}$$

$$F_{b2} = K_b \cdot B_b \cdot B_v \tag{3.33}$$

$$F_{b3} = K_b \cdot B_b \cdot B_v \tag{3.34}$$

$$F_{b4} = K_b \cdot B_b \cdot B_v \tag{3.35}$$

$$F_{b5} = K_b \cdot B_b \cdot B_v \tag{3.36}$$

where K_b represent the film thickness, B_b the strip width and B_v the strip velocity.

The concentration of hydrochloric acid is measured by determining the conductivity of the rinse liquid, since conductivity is a measure of the number of free ions in a liquid, (Norton, 1970). The conductivity is also influenced by other types of ions, which cannot be assigned to hydrochloric acid. This can lead to somewhat higher measured conductivity than the corresponding concentration of hydrochloric acid. From Fig. 3.10, it is seen that at low concentrations of hydrochloric acid, the relation between conductivity and concentration is linear, but at higher concentration of hydrochloric acid the relation is non-linear, (Kematron, 1990). The conductivity is measured in all the rinse tanks. The highest value is achieved in rinse tank 1. Measurement shows that the conductivity is lower than 10 [S/m]. This means that the relation between the concentration of hydrochloric acid and conductivity can be considered to be linear, see Fig. 3.10.

Physically, the transducers are placed in the pipeline which transports the rinse liquid from the rinse rank to the spray equipment. The circulating flow of rinse liquid is taken from the bottom of the rinse tank. Consequently, the conductivity of the rinse liquid is measured from the liquid taken from the lower part of the rinse tank. The degree of conductivity in the rinse tanks is given by:

$$L_i = L_m \cdot C_i \qquad (i=1..5) \qquad (3.37)$$

In equation (3.37), L_m is a calibration constant. From the left-hand diagram of Fig. 3.10, the calibration constant is calculated for hydrochloric acid to be $L_m = 1100$ [mS m^2/kg]. It must be pointed out that the rate of conductivity is dependent on the temperature of the liquid, but temperature compensation is incorporated in the measurement device.

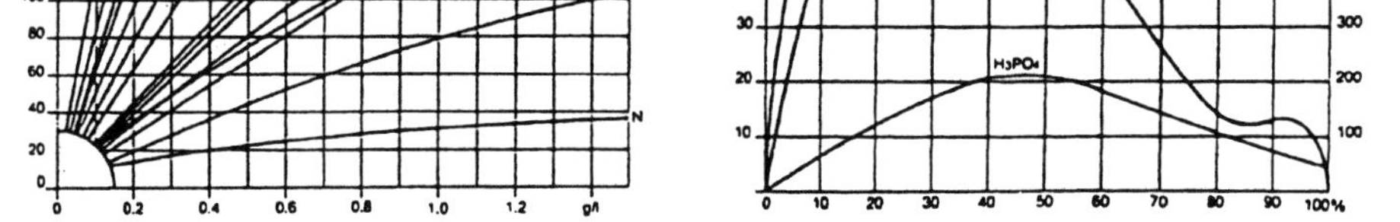

Fig. 3.10 Conductivity curves.

The left-hand diagram in Fig. 3.10 describes the relationship between concentration of a liquid and the corresponding conductivity for weak solutions. The right-hand diagram represents the conductivity for strong acids.

3.2.5 State space model

The basic equations which are formulated in the previous section will be used for identification. For that purpose, we structure the model equations in such a way that they are practical to be handle in connection with the identification software.

It is natural to let the states be the concentration for each of the tanks. This gives five states in our case which are denoted as:

$x_1=C_1$ (Concentration in tank 1)

$x_2=C_2$ (Concentration in tank 2)

$x_3=C_3$ (Concentration in tank 3)

$x_4=C_4$ (Concentration in tank 4)

$x_5=C_5$ (Concentration in tank 5)

The process inputs are the flow of clean water to tank 5 and tank 2. We also have inputs as process parameters: the strip width, the strip velocity and, as will be shown later, the strip thickness. The process parameters vary and are measurable but not possible to use as control inputs. We denote as follows:

$u_1=F_{c1}$ (Flow of clean water into tank 5)

$u_2=F_{c2}$ (Flow of clean water into tank 2)

$u_3=B_b$ (Strip width)

$u_4=B_v$ (Strip velocity)

$u_5=B_t$ (Strip thickness)

The derivatives of the state variables are expressed as functions of the states and the inputs by equation (3.22) - (3.26) and the algebraic equations given by equations (3.27) - (3.31).

The outputs of the process are given by:

$y_1 = L_m \cdot x_1$	(Conductivity measured in tank 1)
$y_2 = L_m \cdot x_2$	(Conductivity measured in tank 1)
$y_3 = L_m \cdot x_3$	(Conductivity measured in tank 1)
$y_4 = L_m \cdot x_4$	(Conductivity measured in tank 1)
$y_5 = L_m \cdot x_5$	(Conductivity measured in tank 1)

We can also introduce internal variables as the flow via the steel strip. These are, however, unknown but they have a physical meaning in model.

F_{b1}	(Flow via the strip into tank 1)
F_{b2}	(Flow via the strip into tank 2)
F_{b3}	(Flow via the strip into tank 3)
F_{b4}	(Flow via the strip into tank 4)
F_{b5}	(Flow via the strip into tank 5)

3.2.6 Model preparation

Discretization

The rinsing process is basically continuous and non-linear. These equations can be summed up as:

$$\frac{dx(t)}{dt} = f\left[x(t),\ u(t),\ s(t)\right] \qquad (3.38)$$

$$y(t) = g\left[x(t)\right] \qquad (3.39)$$

In equations (3.38) and (3.39), f is a non-linear function, $x(t)=[x_1(t)\ x_2(t)\ x_3(t)\ x_4(t)\ x_5(t)]^T$ is the state vector of concentrations and $u(t)=[F_{c1}(t)\ F_{c2}(t)]^T$ is the control vector. Further, $s(t)=[B_b(t)\ B_v(t)\ B_t(t)]^T$ is an additional vector of inputs consisting of strip width, strip velocity and strip thickness. The output transducers are specified by the matrix $g=\text{diag}[L_m, L_m, L_m, L_m, L_m]$.

The non-linear function f contains unknown parameters, which are to be estimated from measured data. The identification software we use in this project needs a process model in a discrete form. Consequently, the model has to be discretized. If the inputs are constant piece-wise, the discretization can be made exactly. However, the strip velocity varies frequently, making an exact discretization impossible. This is also the case with the input flow of clean water, during normal operation of the process. The discrete version of the model is achieved from equation (3.40), (Kailath, 1980). Prerequisite for this method is the model is linear dependent of the states, which is fulfilled for the rinsing process, se equations (3.22) - (3.26).

$$x(t+h) = \Phi(t+h,t) \cdot x(t) \tag{3.40}$$

where $\Phi(t+h)$ is the transition matrix and h is the sampling interval.

Assume that f[x(t),u(t),s(t)] in equation (3.38) varies stepwise during a sampling interval h; this means that f is $f_1 \ldots f_n$ during the intervals $h_1 .. h_n$. Where:

$$\sum_{i=1}^{n} h_i = h \tag{3.41}$$

The transition matrix becomes:

$$\Phi(t+h) = e^{f_1 \cdot h_1} \cdot \;\ldots\ldots\ldots\; \cdot e^{f_n \cdot h_n} \tag{3.42}$$

If the sampling interval is short compared to the time constant of the process and the sub-intervals are short compared to the variations of the function f, the transition matrix can then be approximated by:

$$\Phi(t+h,t) \approx (1 + f_1 \cdot h_1) \cdot \;\ldots\ldots\; (1 + f_n \cdot h_n) \tag{3.43}$$

As $f_1 \cdot h_1 \ldots f_n \cdot h_n$ are small, we can further approximate:

$$\Phi(t+h,t) \approx 1 + f_1 \cdot h_1 + f_2 \cdot h_2 + \;\ldots\ldots\; + f_n \cdot h_n \tag{3.44}$$

This means that the discretized model can be formulated as:

$$x(t+h) = x(t) + \sum_{i=1}^{n} h_i \cdot f_i \cdot x(t) \tag{3.45}$$

From equation (3.45) it is concluded that the mean value of f[x(t),u(t),s(t)] during a sample interval, h, should be used instead of the value of f[x(t),u(t),s(t)] at the sample instant. So we use the mean value of the strip velocity during a sample interval, h. For the same reason, we use the mean value of clean water during normal operation of the process fed into tank 5. In the experiment section, the sample interval will be discussed further.

Scaling

The conductivity rates differ a lot among the five rinse tanks. To avoid numerical problems during identification of unknown parameters, the process outputs are scaled. We use an approximate mean value of the conductivity rate in each rinse tank as scale factors.

Measured value from	tank 1 is scaled with the scale factor:	N_1=3000
	tank 2	N_2=500
	tank 3	N_3=25
	tank 4	N_4=3
	tank 5	N_5=1

Furthermore, the strip velocity has a value lying between 0 - 230 [m/min.] and is scaled with 230. The strip width has a value between 600 - 1600 [mm] and is scaled with 1000. The strip thickness is not scaled because its value does not vary within a large interval. Introduce scaled conductivities $y_1(t)$ to $y_5(t)$ and scaled concentrations $x_1(t)$ to $x_5(t)$. This gives the following discretized state equations and output equations:

$$x_1(t+h) = x_1(t) + k_1 F_{b1} c_0 - k_2 F_{b2} x_1(t) + k_3 F_{m2} x_2(t) - k_4 F_{m1} x_1(t) \tag{3.46}$$

$$x_2(t+h) = x_2(t) + k_5 F_{b2} x_1(t) - k_6 F_{b3} x_2(t) + k_7 F_{m3} x_3(t) - k_8 F_{m2} x_2(t) \tag{3.47}$$

$$x_3(t+h) = x_3(t) + k_9 F_{b3} x_2(t) - k_{10} F_{b4} x_3(t) + k_{11} F_{m4} x_4(t) - k_{12} F_{m3} x_3(t) \tag{3.48}$$

$$x_4(t+h) = x_4(t) + k_{13} F_{b4} x_3(t) - k_{14} F_{b5} x_4(t) + k_{15} F_{m5} x_5(t) - k_{16} F_{m4} x_4(t) \tag{3.49}$$

$$x_5(t+h) = x_5(t) + k_{17} F_{b5} x_4(t) - k_{18} F_{m5} x_5(t) \tag{3.50}$$

$$y_1(t) = x_1(t) \qquad (3.51)$$

$$y_1(t) = x_1(t) \qquad (3.52)$$

$$y_1(t) = x_1(t) \qquad (3.53)$$

$$y_1(t) = x_1(t) \qquad (3.54)$$

$$y_1(t) = x_1(t) \qquad (3.55)$$

In equation (3.46), the concentration on the steel strip at the entrance to the rinsing process is considered as constant, but unknown. Therefore, this concentration is scaled to one and we let the value of the concentration be included in the estimated parameter K_{b1} in equation (3.46).

In equations (3.46) - (3.50) the constants k_1 to k_{18} are computed as:

$$k_1 = \frac{h \cdot L_m}{V_1} \quad k_2 = \frac{h \cdot L_m}{V_1} \quad k_3 = \frac{h \cdot L_m \cdot N_2}{V_1 \cdot N_1} \quad k_4 = \frac{h \cdot L_m}{V_1}$$

$$k_5 = \frac{h \cdot L_m \cdot N_1}{V_2 \cdot N_2} \quad k_6 = \frac{h \cdot L_m}{V_2} \quad k_7 = \frac{h \cdot L_m \cdot N_3}{V_2 \cdot N_2} \quad k_8 = \frac{h \cdot L_m}{V_2}$$

$$k_9 = \frac{h \cdot L_m \cdot N_2}{V_3 \cdot N_3} \quad k_{10} = \frac{h \cdot L_m}{V_3} \quad k_{11} = \frac{h \cdot L_m \cdot N_4}{V_3 \cdot N_3} \quad k_{12} = \frac{h \cdot L_m}{V_3}$$

$$k_{13} = \frac{h \cdot L_m \cdot N_3}{V_4 \cdot N_4} \quad k_{14} = \frac{h \cdot L_m}{V_4} \quad k_{15} = \frac{h \cdot L_m \cdot N_5}{V_4 \cdot N_4} \quad k_{16} = \frac{h \cdot L_m}{V_4}$$

$$k_{17} = \frac{h \cdot L_m \cdot N_4}{V_5 \cdot N_5} \quad k_{18} = \frac{h \cdot L_m}{V_5}$$

Note also that we let the calibration constant of the transducers be included in k_1 to k_{18}.

The discretized and scaled process model can be summed up by the following equations:

$$x(t+h) = F\left[x(t),\ u(t),\ s(t)\right] \tag{3.56}$$

$$y(t) = G\left[x(t)\right] \tag{3.57}$$

where $x(t)=[x_1(t)\ x_2(t)\ x_3(t)\ x_4(t)\ x_5(t)]^T$ is the state vector; $u(t)=[F_{c1}(t)\ F_{c2}(t)]^T$ the control vector; $s(t)=[B_b(t)\ B_v(t)\ B_t(t)]^T$, the additional input vector and $y(t)=[y_1(t)\ y_2(t)\ y_3(t)\ y_4(t)\ y_5(t)]^T$, the output vector. Note that strip velocity and strip width are here scaled in the vector s(t).

Disturbances

To try and make a model structure a perfect copy of the process is not advisable. The effort would be too expensive and the structure of the model would be too complex. By studying the prediction error of the deterministic model given by equation (3.56) and (3.57), it is possible to investigate whether there are unmodelled parts of the process. Unmodelled parts can be detected by the fact that the prediction error is not white. Unmodelled parts of the process may originate from:

- **Incomplete mixing.** It has been assumed that perfect mixing occurs in the rinse tanks. It is possible that the upper part of the tank volume has a higher concentration than the other part of the tank. This is because the rinse water sprayed at the steel strip lands in the upper part of the tank volume. The thickness of this layer may also be dependent on the value of the main flow.

- **Incomplete rinsing.** It is possible that the steel strip may be incompletely rinsed during high strip velocity.

- **Evaporation from the rinse zones.** The evaporation which in fact occurs has an additive influence on the main flow. For instance, the main flow from tank 2 is reduced by the sum of evaporation from tank 2 to tank 5.

- **Dynamic main flow.** Simulation of the main flow shows that the rise time of rinse tank 2 is longer than the rise time of tank 5. Furthermore, the knowledge about the main flow from tank 5 is better than from tank 2 because the flow into tank 5 is known.

- **Varying level in rinse tank 1.** The level of liquid in rinse tank 1 varies around a reference value. This means that the volume in tank 1 is not constant, contrary to what is assumed in equation (3.22).

In addition to simplifications in the structure of the model, the process is influenced by disturbances from the following sources:

- **Varying concentration of hydrochloric acid in the last pickling tank.** Hydrochloric acid is consumed during pickling of the steel strip. This consumption depends on production variables such as strip velocity, strip width and strip thickness. Furthermore, the quality of the strip surface influences the consumption of hydrochloric acid. The concentration of hydrochloric acid in the pickling tanks is controlled by a separate controller, but the concentration is not constant.

- **Buckled steel strip.** Sometimes it happens that a buckled steel strip passes the rinsing process. The edge or the middle of the strip may be buckled. An edge-buckled steel strip occurs during hot rolling, when the edges of the steel strip are reduced more than the middle of the strip. A middle-buckled steel strip occurs, when the middle of the strip is reduced more than the edges. The buckled strip may cause problems for the squeezer rolls to reduce the flow via the strip.

- **Too much pickling.** During longer stops, the hydrochloric acid is drained off the pickling line, because the steel strip should not be in contact with the acid too long. Otherwise, there is a risk of too much pickling. Hydrochloric acid can however be left on the steel strip when the pickling tanks are drained off and acid is forced deeper into the steel surface. This means that it is difficult to rinse this part of the steel strip.

- **Rust on the steel strip.** During longer stops, the steel strip rusts in those areas of the steel strip which are in contact with the squeezer rolls. These areas will be difficult to rinse and more hydrochloric acid is transferred further through the rinsing process.

Unmodelled parts of the process and other disturbances are added as disturbances to the state of the process. During identification, these disturbances are regarded as white noise. This is also a simplification but colouring of the state noise needs extra states, which is time-consuming during identification.

Measurement of conductivity is also influenced by disturbances. These disturbances are also regarded as white noise. From specification of the conductivity transducers the noise is estimated to be 1%, (Kematron, 1990).

3.2.7 General form

Equations (3.56) and (3.57) are extended to include process noise on all states and measurement noise on all outputs. This gives the following discrete model of the rinsing process:

$$x(t+h) = F\left[x(t),\ u(t),\ s(t)\right] + v(t) \tag{3.58}$$

$$y(t) = G\left[x(t)\right] + w(t) \tag{3.59}$$

It is assumed that the process noise, $v(t)=[v_1(t)\ v_2(t)\ v_3(t)\ v_4(t)\ v_5(t)]^T$, and measurement noise, $w(t)=[w_1(t)\ w_2(t)\ w_3(t)\ w_4(t)\ w_5(t)]^T$, are independent white noise. Consequently the covariance matrices of v(t) and w(t) are diagonal matrices. The covariance matrices for v(t) and w(t) can be written as $\mathrm{cov}[v(t)]=R_1$ and $\mathrm{cov}[w(t)]=R_2$, where R_1 and R_2 are diagonal matrices.

3.3 Summary

A basic model of the steel strip rinsing process is formulated and is based on hypotheses, facts and assumptions. Knowledge about the process is incomplete and leads to a trial model. The result derived from the basic modelling is not certain and must be adapted to measured data from the process.

During the basic modelling phase the process is divided into subprocesses. In our case is each rinse zone is represented by a submodel. These parts are connected by two different flows, one flow via the strip and one main flow of rinse liquid between the rinse tanks. These flows go in opposite directions. The flow originating from the evaporation is considered as small compared to the other flows and therefore is ignored.

The basic model is based on the principle of mass balance. During modelling, several assumptions have been made to model only the physical phenomena which are important to the function of the process. These assumption can be summarised as follows:

- Evaporation from the rinse zone is small compared to the main flow.
- The main flow is a stiff system. Simulation of the main flow indicates a rise time of about 5 minutes, which is very short compared to the dominating time constants of the system.
- The process is a mixing process of different flows with different concentrations of hydrochloric acid. No chemical reaction takes place.
- The mixing in the rinse tanks is perfect. It takes about five minutes to turn over the volume of the tanks by means of the circulating rinse flow.
- The steel strip is well rinsed when it has passed a rinse zone. This means that the liquid of the film at the end of the rinse zone has the same concentration as the liquid in the rinse tanks.

The amount of the flow via the steel strip is a result of the efficiency of the squeezer rolls. The basic model is suitable to supervise the flow via the steel strip, because this flow is modelled explicitly, separately from other parts of the process. The model is also suitable to control the process, because the model is in a state space form and describes a dynamic relation between inputs, states and process outputs.

4 APPLICATION OF AUGMENTED MODELLING

4.1 Introduction

The basic model of the process, achieved in Chapter 3, is uncertain and incomplete. Therefore, the basic model has to be expanded and adapted to measured data. This procedure follows the method given in Chapter 2 and involves experimentation, expanded modelling, parameter identification, model analysis and model appraisal.

The model expansion is based on model analysis and the interpretation of the analysis. This is not only a matter of science but also of craft. This means, that the model builder must be able to combine expertise knowledge and intuition. It is very difficult to formulate this intuition in words so this presentation of the expended modelling and identification is mainly focused on skill.

In Section 4.2, we discuss the performance of the experiment with the process to achieve informative measured data. Importance parameters are; sample time, duration and amplitude of the control signal. In Section 4.3, we search for an appropriate model. This involves also the purpose of the model. The chapter is concluded with other possible model attempt and a summary.

4.2 Experiment

During the experiment with the process, the flow of clean water fed into tank 5 is controlled by a PRBS signal and the normal controller is closed. The flow is instead a signal which is constant piece-wise with a high level at 3.0 [m^3/h] and low level at 1.0 [m^3/h]. The high level of the flow is limited by the production of clean water and the low level is based on the fact that the rinse water in tank 5 must be clean enough to achieve a well-rinsed steel strip.

The experiment is performed twice with the same input sequence. The second experiment is done to obtain data which can be used for cross validation of the process model. In the following, the measurement sequences are denoted by sequence 1 and sequence 2. When the process is running, there are shorter stops occasionally. During these stops, the flow of clean water is shut down to zero.

It is not possible to influence the other inputs: strip velocity, strip width and strip thickness, since they depend on the production planning and the actual volume of orders.

Besides the control sequence of clean water fed into tank 5, we have to decide the size of the sample time, h. It is related to the dominating time constant of the process, the variation of the production parameters and also the discretization method. In our case, the concentrations in the tanks vary slowly but the strip velocity varies frequently. The discretization method is based on Euler's method, which requires a relatively short sample time.

A reasonable sample time relatively the dominating time constant and Euler's method is to let h=0.2 [h]. Since the strip velocity varies often during 0.2 [h], we need to sample this production parameter with a higher frequency. After studying the strip velocity a reasonable sample time for this parameter is 0.01 [h]. The mean value is computed for the speed during 0.2 [h] and is used as a value of the strip velocity. This means an extra uncertain factor, which is included in the model. This kind of approximation is discussed further in section 3.2.6.

The results of the first sequence of the experiment is shown in Fig. 4.1 and Fig. 4.2. The second measurement sequence has the same time duration and the variations in the signals are about the same as for sequence 1.

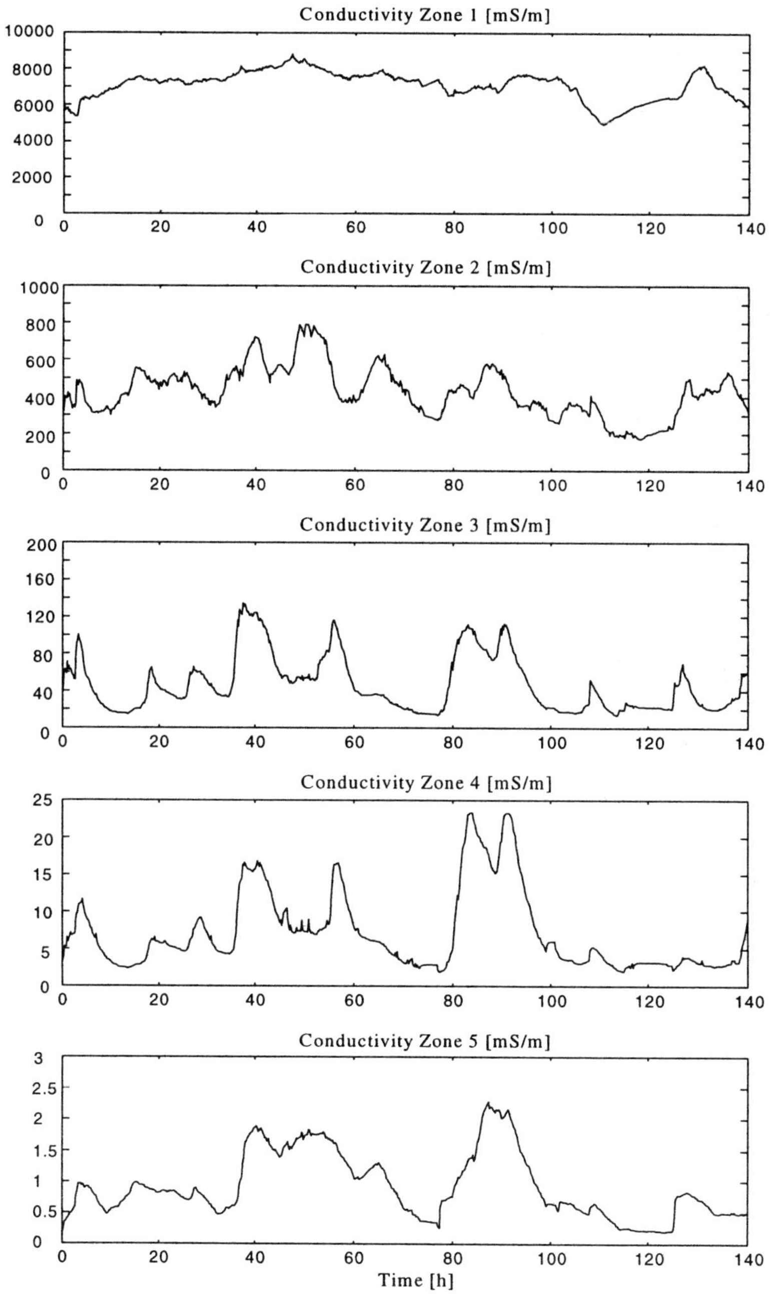

Fig. 4.1 Process outputs, sequence 1.

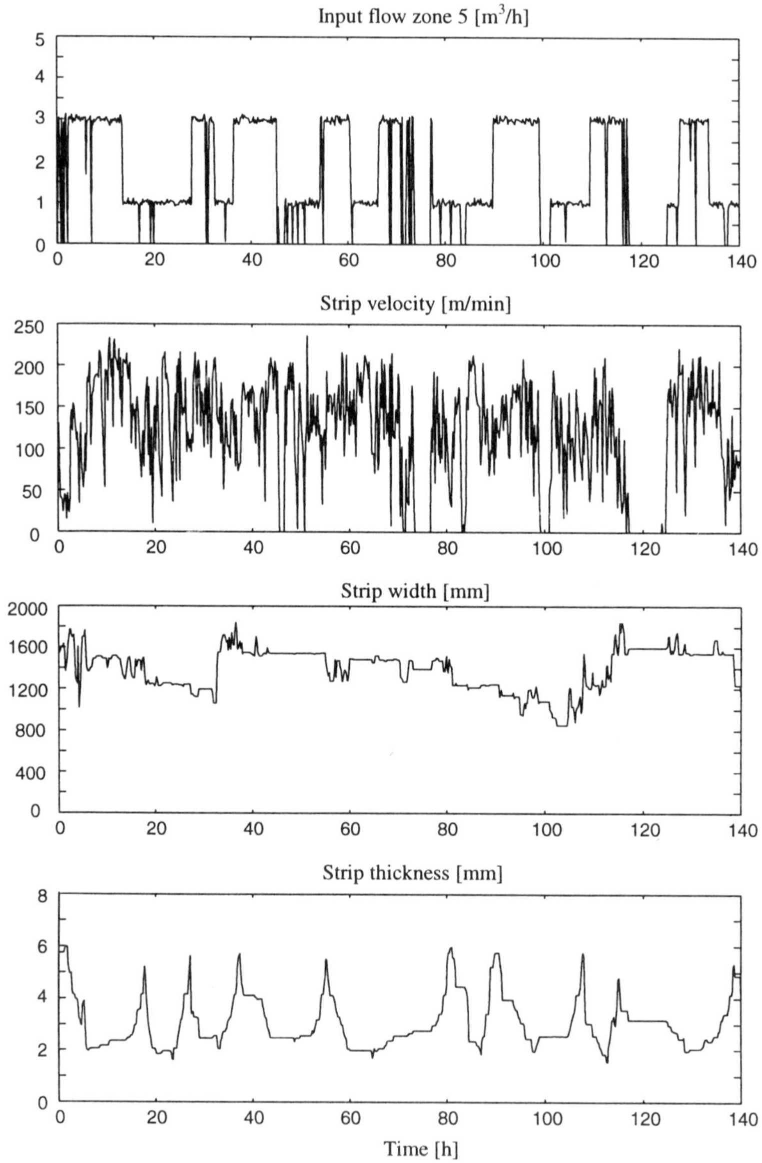

Fig. 4.2 Process inputs, sequence 1.

❐ Comments on the measurements

Output. The conductivity in rinse tank 1 does not vary so much. This is natural, since the dilution flow from this tank is only 0.3 [m^3/h]. The conductivity is a at relatively high level during the experiment; measurements from control experiments with the process, presented in Chapter 6, show a conductivity measurement of about 3000 [mS/m].

Measurements in the other rinse tanks show that the amplitude of the conductivity varies a great deal. This suggests that these signals are "informative" and can be used for model identification. We should also note that the concentrations between two nearby rinse tanks differ by factor of about 10.

Input. The flow of clean water fed into tank 5 varies in accordance with the desired form, between 1 - 3 [m^3/h]. Note, that the amplitude of flow shows small variations, since the presented signal arising from the measurements of the actual flow.

The strip velocity varies frequently. This is typical of the way the process is operated and depends mainly on the fact that it takes time to weld the ends of two coils at the entrance to the process. It also takes time to split the steel strip into coils again at the exit from the process. The entry and delivery loop cars cannot absorb all the delay. The strip velocity varies between 0 - 240 [m/min.].

The strip thickness has a typical variation between thin and thick steel strip. The strip thickness varies between 2 - 6 [mm].

The strip width also has a typical variation between narrow and broad steel strip. The strip width varies between 600 -1800 [mm].

4.3 Searching for an appropriate model

In Chapter 3, we have concluded that the basic model contains of unknown parts. For the rinsing process these consist of the flow via the steel strip, directly after the squeezer rolls. A first attempt to describe this flow is to make a sub-model proportional to the steel area passed per time unit.

The estimation of unknown parameters is based on the Likelihood function. We distinguish between two cases, which are called the deterministic and the stochastic model. The deterministic model is achieved without updating the process states by using an Extended Kalman Filter while the stochastic model is attained by updating the states by using a filter.

Measurements of the conductivity is affected by noise. We suppose that the noise is zero mean Gaussian white noise with an approximate standard deviation of about 1%. Since the process outputs are normalised with the corresponding mean value, the measurement covariance matrix is set to:

$$R_2=\mathrm{diag}[1\cdot 10^{-4}\ \ 1\cdot 10^{-4}\ \ 1\cdot 10^{-4}\ \ 1\cdot 10^{-4}\ \ 1\cdot 10^{4}]$$

The states are also corrupted by noise, which corresponding covariance matrix is used by the Extended Kalman Filter updating the states. In our case the state covariance matrix is set to:

$$R_1=\mathrm{diag}[1\cdot 10^{-3}\ \ 1\cdot 10^{-3}\ \ 1\cdot 10^{-3}\ \ 1\cdot 10^{-3}\ \ 1\cdot 10^{-3}]$$

This matrix is achieved by searching for an approximation of lowest value of the Likelihood function. This is done by manually changing the elements of the covariance matrix. For each attempt, the search for an appropriate model includes the following; expanded modelling, identification, model analysis and model appraisal.

4.3.1 Basic model $\mathcal{M}_1$

❒ **Basic modelling**

The basic model is developed in section 3.2 and we assume that the film thickness after each pair of squeezer rolls is equal. The flow via the steel strip after rinse zone 5 becomes very small. This is explained by the fact that after rinse zone 5, the steel strip passes between a double pair of rolls, see Fig. 3.3. Therefore, in the model we let $F_{b6}=0$, see also Fig. 3.6.

The steel strip also passes between a double pair of squeezer rolls at the entry of the rinsing process. The flow via the strip at the entry of the process cannot be set to equal zero, because the concentration of this flow is very high compared to the concentration on the strip after rinse zone 5.

From equations (3.32) - (3.36), we have modelled of the unknown parts as:

$$F_{bi}=K_b \cdot B_b \cdot B_v \qquad (i=1\ldots5) \qquad (4.1)$$

where K_b is the film thickness, B_b strip width and B_v strip velocity.

❒ **Identification**

The basic model contains only one unknown parameter. The results from the identification are presented in table 4.1, where "deterministic" and "stochastic" means that the identification is done without a Kalman filter and with a filter respectively.

Table 4.1 Estimation results model $\mathcal{M}_1$.

	Deterministic	*Stochastic*
Loss function	$54\ 574\ 10^3$	19 222
K_b	0.2884	0.2484

❒ **Model analysis**

The value of the estimated parameter, K_b, is about the same for both deterministic and stochastic identification. This is promising, since the parameter estimation works also when the states are updated via the Extended Kalman Filter. The value of the loss function is used later for model comparison.

At this early stage, it is a good idea to plot the measured process output and the output from the ballistic simulation in the same diagram. It give a quick glimpse of the behaviour of the model, see Fig. 4.3.

The ballistic simulation shows that the conductivity in tank 1 is very low compared with the measured. We can also see that the simulated conductivity in the tank 2 to tank 5 varies, but the quality of these variations is not good enough.

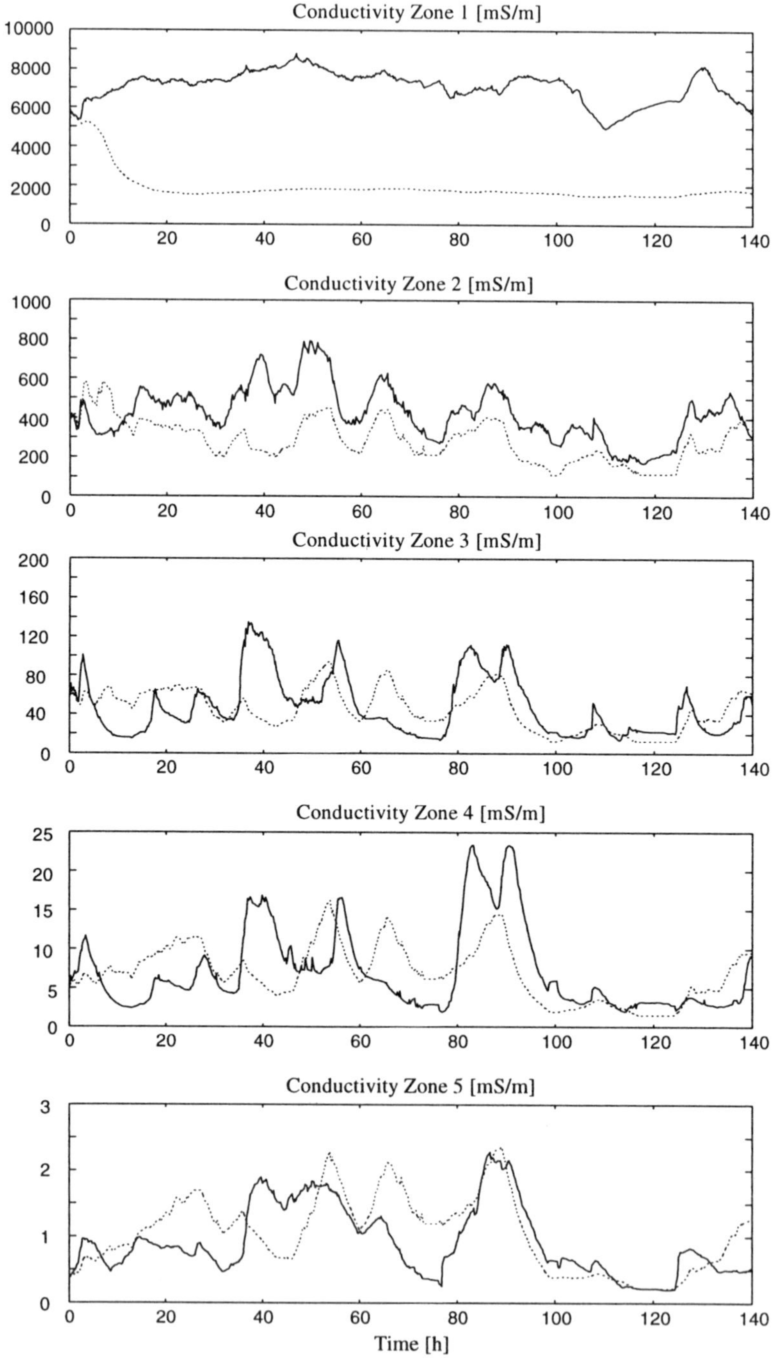

Fig. 4.3 Ballistic simulation with model $\mathcal{M}_1$, *dotted* - model output.

❐ **Model appraisal**

The same parameter is used to estimate the thickness of the liquid carried by after each pair of squeezer rolls. The estimate will be a compromise for the real liquid thickness after each pair of rolls and result in incorrect steady state levels of the conductivity. This is obvious for the degree of conductivity in tank 1.

4.3.2 Expanded model $\mathcal{M}_{121}$

❐ **Expanded modelling**

The continuing of the expanded modelling procedure seems obvious: the parameter K_b in equation (4.1) is divided into two parameters, one parameter for squeezer rolls 1 and one parameter for the other squeezer rolls. This gives the flows via the steel strip is modelled as:

$$F_{b1}=K_{b1}\cdot B_b\cdot B_v \qquad (4.2)$$

$$F_{bi}=K_{b2}\cdot B_b\cdot B_v \qquad (i=2\ldots5) \qquad (4.3)$$

The basic model $\mathcal{M}_1$ is expanded to model $\mathcal{M}_{121}$. The notation of $\mathcal{M}_{121}$ means; the model originate from the basic model $\mathcal{M}_1$ and is expanded to the level 2 and at this level we are dealing with model number 1.

❐ **Identification**

Two unknown parameters are identified and the results are presented in table 4.2.

Table 4.2 Estimation results model $\mathcal{M}_{121}$.

	Deterministic	*Stochastic*
Loss function	39 791 10^3	18 610
K_{b1}	1.3367	1.5218
K_{b2}	0.2352	0.2507

❒ **Model analysis**

The values of the estimated parameters differ a lot; the parameter K_{b1} is increased and the parameter K_{b2} is decreased compared to the estimate with only one parameter. This is a result of a better adaptation to the measured data. We can also see that the values of the estimated parameters from the deterministic and stochastic models are about the same size.

❒ **Model appraisal**

The loss function decreases considerably compared to the basic model $\mathcal{M}_1$. This indicates that our decision to divide the parameter K_b into two variables is a good idea. So far, the way of expanding the model seems promising. There is no reason to stop here, instead we ought to go on in the started way.

4.3.3 Expanded model $\mathcal{M}_{122}$

❒ **Expanded modelling**

The loss function decreases considerably when expanding the basic model $\mathcal{M}_1$ to model $\mathcal{M}_{121}$. Keeping in mind that the squeezer rolls are replaced when the process operator considers a pair is worn, we continue the expanded modelling operation with letting the flow after each pair of squeezer rolls have an individual performance. The flows via the steel strip are accordingly modelled as:

$$F_{bi}=K_{bi}\cdot B_b\cdot B_v \qquad (i=1\ldots5) \tag{4.4}$$

❒ **Identification**

Five parameters are estimated and the results is presented in table 4.3.

Table 4.3 Estimation results model $\mathcal{M}_{122}$.

	Deterministic	*Stochastic*
Loss function	$31\ 510\ 10^3$	14 310
K_{b1}	1.2900	1.2391
K_{b2}	0.1047	0.1453
K_{b3}	0.3132	0.2578
K_{b4}	0.4059	0.4147
K_{b5}	0.3673	0.2767

❒ Model analysis

From table 4.3, we can see that the difference in magnitude of the estimated parameters justifies the idea that the squeezer rolls have individual behaviour.

The ballistic simulation results are presented in Fig. 4.4. We can see that the simulated conductivity in tank 1 and tank 2 follows the dynamics of the measurements, but the simulated conductivity in the others is not able to describe the dynamics of the process. The stochastic simulation results, the residuals, are presented in Fig. 4.5. We can see that only the residuals achieved from tank 1 seemed to be a white noise process. The other sequences of residuals include some kind of low frequency noise. This can also be seen from Fig. 4.6, which presents the auto-correlation of the residuals from all five outputs.

Since the innovation process contains some kind of retain information, we have to analyse this further. It is not at all sure that we can trace any information in the innovation process, but one way to proceed with the investigation is to make a cross-correlation test between the residuals and the inputs. In Fig. 4.7, the cross-correlation between the residuals and the strip thickness is presented.

From Fig. 4.7, we can see that there is retained information concerning the influence of the strip thickness, specially, in zone 3 and zone 4. This means that the influence of the strip thickness should be incorporated in the model. It can indicate that there is a flow beside the squeezer rolls, which is due to the strip thickness.

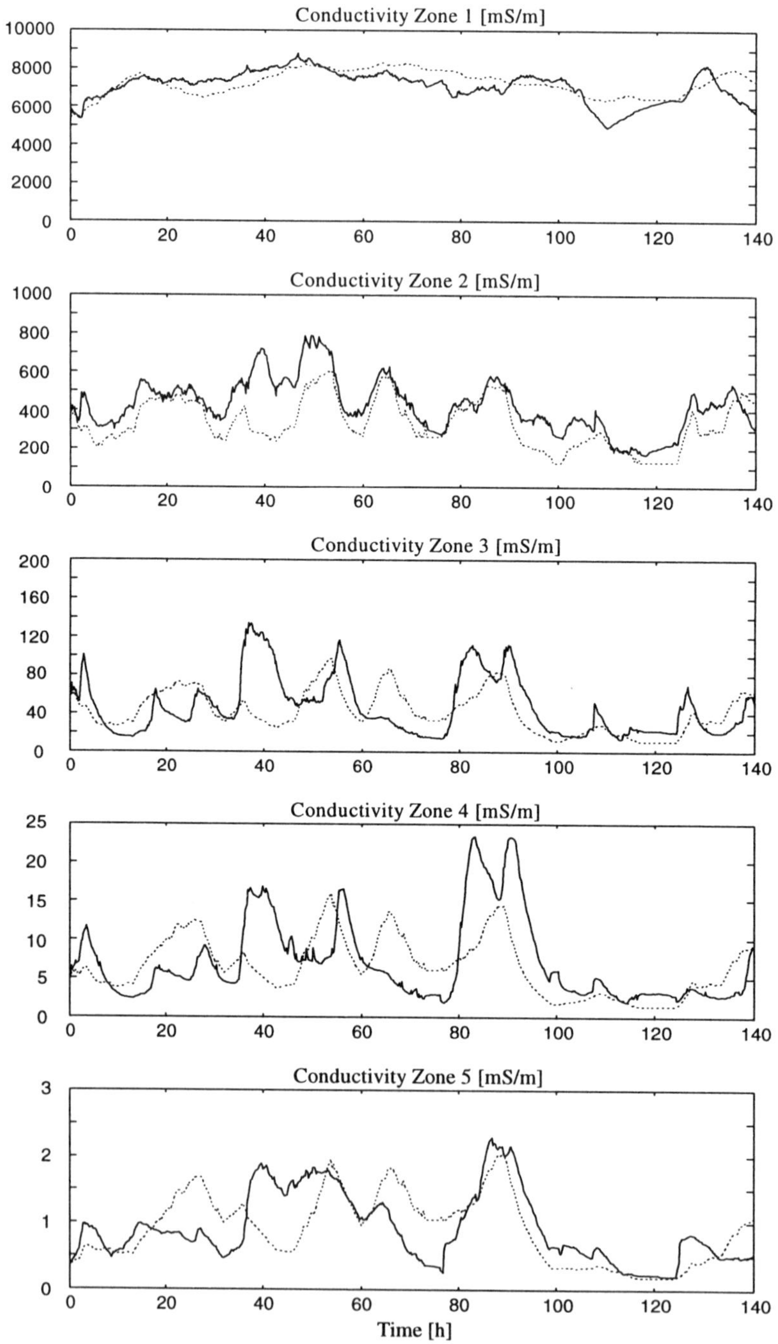

Fig. 4.4 Ballistic simulation with model $\mathcal{M}_{122}$, *dotted* - model output.

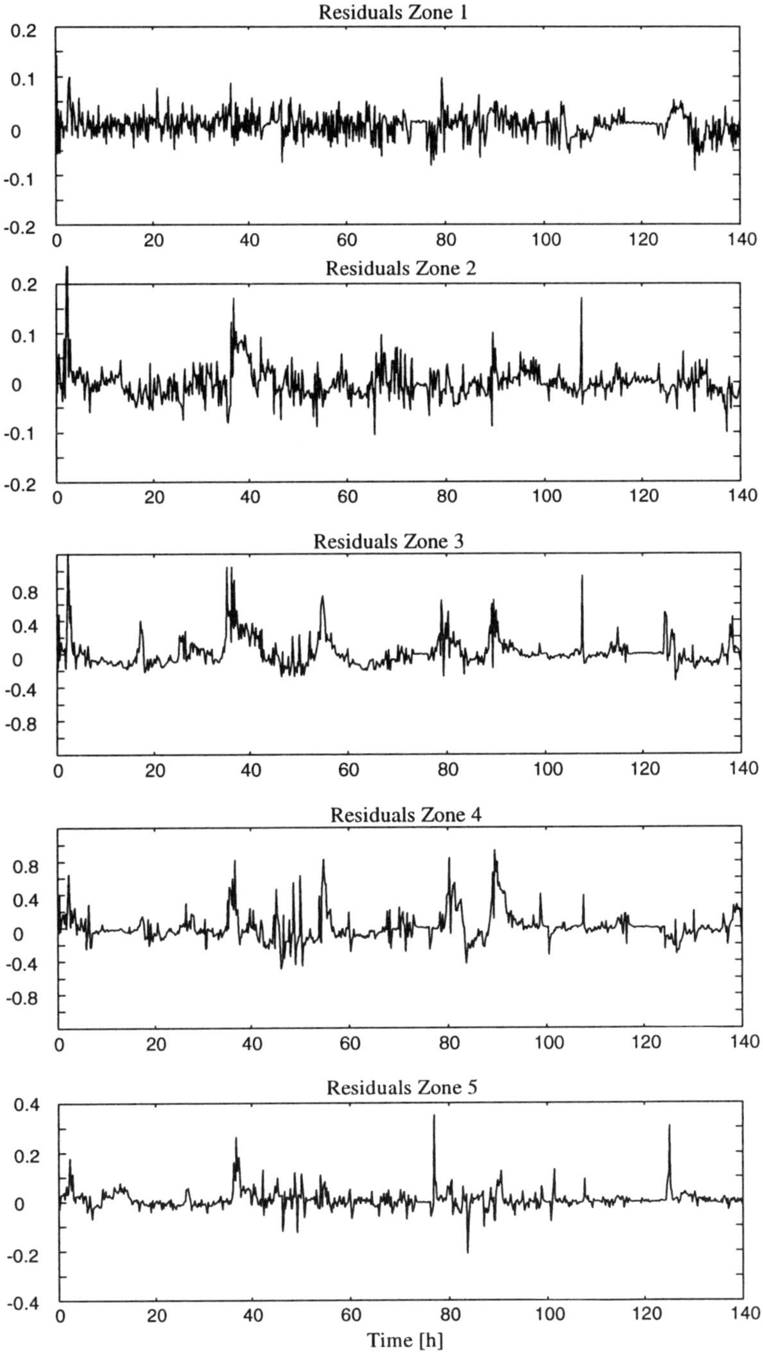

Fig. 4.5 Residual from stochastic simulation with model $\mathcal{M}_{122}$.

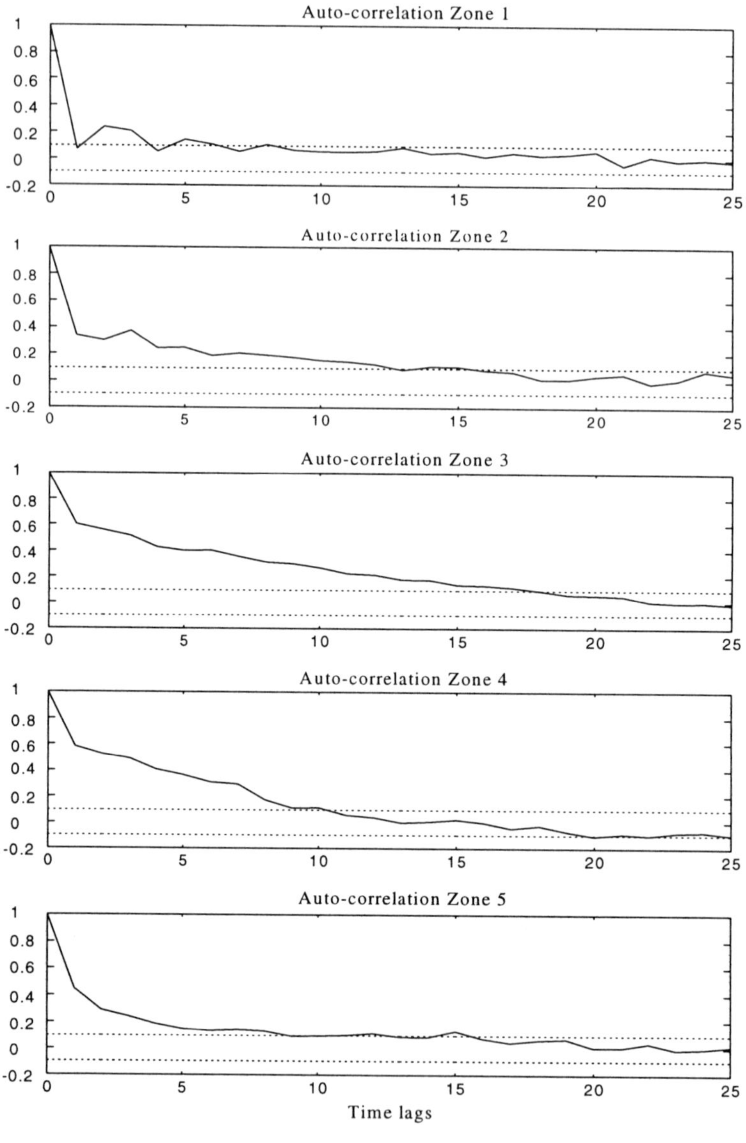

Fig. 4.6 Auto-correlation model $\mathcal{M}_{122}$, *dotted* 99% confidence interval.

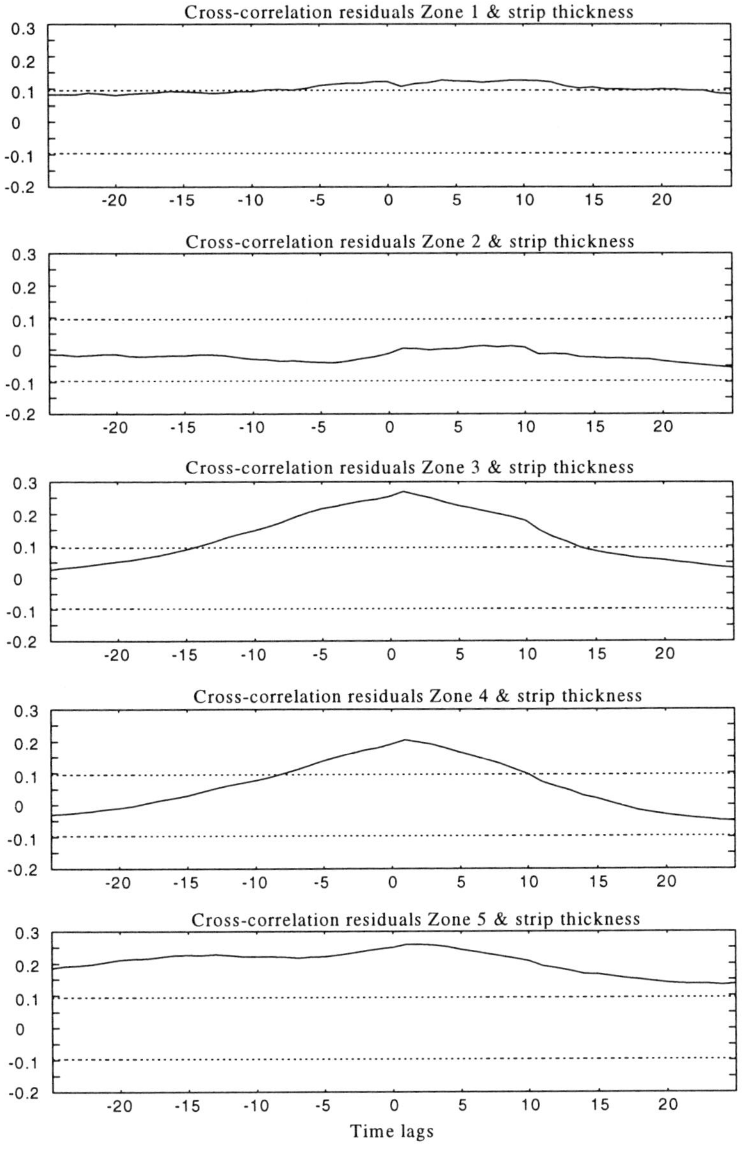

Fig. 4.7 Cross-correlation between the residuals and the strip thickness, model $\mathcal{M}_{122}$, *dotted* 99% confidence interval.

❐ **Model appraisal**

The loss function decreases significantly compared to model $\mathcal{M}_{121}$. This indicates again that our decision to increase the number of estimated parameters is a good idea. The ballistic simulation of model $\mathcal{M}_{122}$, see Fig 4.4, is improved compared to the basic model $\mathcal{M}_{1}$. Both the decrease in loss function and the results from simulation favour the model $\mathcal{M}_{122}$.

The results of the model analysis have given a surprising indication: the cross-correlation shows that the process is dependent on the strip thickness. A reasonable assumption is that there is a flow beside the steel strip which must be incorporated in the model. This finding is new process knowledge.

4.3.4 Expanded model $\mathcal{M}_{131}$

❐ **Expanded modelling**

Taking the indication that the squeezer rolls have an individual performance for granted, we expand the model of the flow via the strip with a strip thickness dependent part. We assume that there is flow beside the strip which is carried by the sides of the steel strip; this gives that an additional flow which is proportional to the strip thickness and the strip velocity. The flow is modelled by equation (4.5).

$$F_{bi}=K_{bi}\cdot B_b\cdot B_v+K_{ti}\cdot B_t\cdot B_v \qquad (i=1...5) \qquad (4.5)$$

❐ **Identification**

Parameter estimation gives negative values for K_{b2}, K_{b3} and K_{b4}. Since the estimated parameters represent the film thickness of the liquid which is carried with the strip, the estimated values must be positive. Furthermore, the loss function is insensitive due to variation in parameter K_{t5}.

❐ **Model appraisal**

The model is rejected, due to unallowable values of estimated parameters.

4.3.5 Expanded model $\mathcal{M}_{132}$

❐ **Expanded modelling**

We reduce the influence of the strip thickness in the model by taking the square root of the thickness. This is a trial model to achieve positive values of the parameters K_{bi}.

$$F_{bi}=K_{bi}\cdot B_b\cdot B_v+K_{ti}\cdot\sqrt{B_t}\cdot B_v \qquad (i=1...5) \qquad (4.6)$$

❐ **Identification**

Estimation of the unknown parameters still gives negative values.

❐ **Model appraisal**

The model is rejected, due to unallowable values of estimated parameters.

4.3.6 Expanded model $\mathcal{M}_{133}$

❐ **Expanded modelling**

It may be that we are following a misleading track. The flow beside the strip may be proportional only to the opening between the squeezer rolls, since the flow beside the strip may originate from that the nozzles, which spray rinse liquid on the strip, are directed towards the squeezer rolls instead of towards the steel strip. This gives a model trial as:

$$F_{bi}=K_{bi}\cdot B_b\cdot B_v+K_{ti}\cdot B_t \qquad (i=1...5) \qquad (4.7)$$

❐ **Identification**

Estimation of the unknown parameters give still negative values.

❐ **Model appraisal**

The model is rejected, due to negative values of estimated parameters, K_{bi}.

4.3.7 Expanded model $\mathcal{M}_{134}$

❐ **Expanded modelling**

At this point we have achieved information that the strip thickness influences the behaviour of the process. However, we have not succeeded in making an appropriate model of the influence. We have to look for a hint how to proceed to achieve an appropriate model. The process operator has a deep knowledge about running the process. This source of information must not be underestimated.

When discussing the concrete problem about the influence of the strip thickness with the process operators, we are told that the rubber of the squeezer rolls can behave in different ways. The rubber can be applied by the manufacturer in various ways and the rubber can be of different composition. There is also sometimes an imprint of the strip on the rubber of the squeezer rolls. Based on this discussion, we assume that the strip thickness must exceed a lower limit before there is a flow beside the strip, and an attempt to model this property is given by:

$$F_{bi}=K_{bi}\cdot B_b\cdot B_v+K_{ti}\cdot(B_t-B_{toffi})\cdot\alpha_i \qquad (i=1\,..\,5) \qquad (4.8)$$

where:

$\alpha_i=1$ when $B_t > B_{toffi}$ else $\alpha_i=0$

The second term in equation (4.8) is explained as follows; when the strip thickness, Bt, is greater than a parameter, B_{toffi}, there will be an opening between a pair of squeezer rolls besides the strip. The size of this opening is modelled as proportional to (B_t-B_{toffi}).

The parameter K_{ti} is set to 1 during the identification, to avoid too many unknown parameters. This is done just to investigate whether or not the proposed model can give positive values of the parameters K_{bi}.

Identification

The results of the estimation of model $\mathcal{M}_{134}$ are presented in table 4.4.

Table 4.4 Estimation results, model $\mathcal{M}_{134}$.

	Deterministic	*Stochastic*
Loss function	13 811 10^3	7 463
K_{b1}	1.0724	1.2384
K_{b2}	0.0522	0.1451
K_{b3}	0.0547	0.2287
K_{b4}	0.3880	0.3661
K_{b5}	0.3563	0.2857
B_{toff2}	6.8634	6.8600
B_{toff3}	2.3030	4.7634
B_{toff4}	4.9950	5.0080
B_{toff5}	6.3604	6.3604

Model analysis

All estimated parameters are positive. The parameters B_{toffi} have a physical meaning which says that the strip thickness must exceed this value before there will be a flow beside the strip; therefore $B_{toffi} \in [2, 7]$, the interval of strip thickness. The loss function decreases significantly compared to model $\mathcal{M}_{122}$.

Model appraisal

The results of the model analysis indicate that the model $\mathcal{M}_{134}$ may be a promising way to model the influence of the strip thickness, but must be further investigated.

4.3.8 Expanded model $\mathcal{M}_{141}$

❐ **Expanded modelling**

We continue to believe that the strip thickness must exceed a lower limit before there is a flow beside the strip and assume that the flow beside the strip is proportional to the opening between the rolls multiplied by the strip velocity. We assume that the flow beside the strip is carried by the side of the strip or by the rolls. Therefore, the second term in equation (4.8) is multiplied with the strip velocity and gives:

$$F_{bi}=K_{bi}\cdot B_b\cdot B_v+K_{ti}\cdot(B_t-B_{toffi})\cdot B_v\cdot\alpha_i \qquad (i=1\ ..\ 5) \qquad (4.9)$$

where:

$\alpha_i=1$ when $B_t > B_{toffi}$ else $\alpha_i=0$

We increase the model level, since we have found a promising model on level 3, which maybe can be expanded further. The parameter K_{ti} is set to 1 during the identification to avoid too many unknown parameters.

❐ **Identification**

The results of the identification of model $\mathcal{M}_{141}$ are presented in table 4.4. The values of the parameters B_{toff2} and B_{toff5} will approach the upper limit and therefore are set equal the limit.

Table 4.5 Estimation results, model $\mathcal{M}_{141}$.

	Deterministic	*Stochastic*
Loss function	11 108 10^3	-3 971
K_{b1}	1.1645	1.2375
K_{b2}	0.0790	0.1449
K_{b3}	0.1504	0.1231
K_{b4}	0.2749	0.2108
K_{b5}	0.3552	0.2988
B_{toff2}	7.0000	7.0000
B_{toff3}	2.6299	3.3304
B_{toff4}	3.5094	3.7236
B_{toff5}	7.0000	7.0000

❐ **Model analysis**

The estimates from the deterministic and stochastic models are about the same. The values are also admissible.

❐ **Model appraisal**

The loss function decreases considerably compared to model $\mathcal{M}_{134}$. This means that model $\mathcal{M}_{141}$ gives a better description of the flow via the strip than model $\mathcal{M}_{134}$.

4.3.9 Expanded model $\mathcal{M}_{142}$

❐ **Expanded modelling**

The model $\mathcal{M}_{141}$ seems to be very promising. We expand the model by searching also on the parameter $K_{ti,}$ and letting this be equal for all squeezer rolls.

$$F_{bi}=K_{bi}\cdot B_b\cdot B_v+K_t\cdot(B_t-B_{toffi})\cdot B_v\cdot\alpha_i \qquad (i=1...5) \qquad (4.10)$$

$\alpha_i=1$ when $B_t > B_{toffi}$ else $\alpha_i=0$

This model is named $\mathcal{M}_{142}$.

❐ **Identification**

The results from the identification of model $\mathcal{M}_{142}$ are presented in table 4.6. The table is enlarged with the diagonal elements of the inverse of the Hessian matrix. These elements are associated with the accuracy of the individual estimates and are used to investigate the identifiability of the estimated parameters. Only the stochastic variant is considered.

For this case, the parameter B_{toff2} is approach the upper limit and is set to the limit.

Table 4.6 Estimation results model $\mathcal{M}_{142}$.

	Stochastic	*Inverse Hessian*
Loss function	-5 122	
K_{b1}	1.2379	$2609.4\ 10^{-6}$
K_{b2}	0.1448	$7.3\ 10^{-6}$
K_{b3}	0.1081	$9.1\ 10^{-6}$
K_{b4}	0.1913	$26.0\ 10^{-6}$
K_{b5}	0.2898	$84.3\ 10^{-6}$
B_{toff2}	7.0000	-
B_{toff3}	2.8808	$207.5\ 10^{-6}$
B_{toff4}	3.3691	$344.1\ 10^{-6}$
B_{toff5}	5.3380	$4444.1\ 10^{-6}$
K_t	0.6846	$56.9\ 10^{-6}$

❐ **Model analysis**

It is not possible to achieve any inverse Hessian related to parameter B_{toff2}, since the corresponding estimate is set equal to the upper limit. The diagonal elements of the inverse Hessian shows the relations for the lower bounds of the variance of the estimates respective. We see that the variance of the estimate of the parameters K_{b1} and B_{toff5} are larger than the others.

❐ **Model appraisal**

The loss function decreases compared to model $\mathcal{M}_{141}$. The degree of freedom is increased by two compared with the model $\mathcal{M}_{141}$ and the loss function is decreased by about 150. From the chi-2 table, this is a significant improvement of the model. The large value of the lower bound of the variance of parameter B_{toff5} is discussed further in Chapter 5. The parameter K_{b1} is necessary for the model structure, because this parameter had to be a part of the model.

4.3.10 Expanded model $\mathcal{M}_{143}$

❒ **Expanded modelling**

The model $\mathcal{M}_{142}$ is expanded by searching on every parameter $K_{ti,}$. It is a natural expansion to investigate in what way the strip thickness has an individual influence on the subprocesses. This model is named $\mathcal{M}_{143}$.

$$F_{bi}=K_{bi}\cdot B_b\cdot B_v+K_{ti}\cdot(B_t-B_{toffi})\cdot B_v\cdot\alpha_i \qquad (i=1...5) \qquad (4.11)$$

$\alpha_i=1$ when $Bt > Btoff_i$ else $\alpha_i=0$

❒ **Identification**

The results of the estimation of model $\mathcal{M}_{143}$ are presented in table 4.7. Note the parameter K_{t2} is not used during the identification phase, since the parameter B_{toff2} is set to a value which rejected influence of the steel strip thickness. This behaviour of the process is also indicated in Fig. 4.7, where the cross-correlation between the residuals from zone 2 and the strip thickness is within the 99% confidence interval.

Table 4.7 Estimation results model $\mathcal{M}_{143}$.

	Stochastic	*Inverse Hessian*
Loss function	-5 126	
K_{b1}	1.2392	2561.3 10^{-6}
K_{b2}	0.1448	7.3 10^{-6}
K_{b3}	0.1087	8.9 10^{-6}
K_{b4}	0.1913	22.9 10^{-6}
K_{b5}	0.2900	85.2 10^{-6}
B_{toff2}	7.0000	-
B_{toff3}	2.8790	189.5 10^{-6}
B_{toff4}	3.3805	429.5 10^{-6}
B_{toff5}	5.1359	3681.9 10^{-6}
K_{t2}	0	-
K_{t3}	0.6809	71.0 10^{-6}
K_{t4}	0.6929	139.0 10^{-6}
K_{t5}	0.4062	10062.1 10^{-6}

❐ **Model analysis**

The estimates are feasible but the difference between the lower bound of the variance is relatively large. This means that the identifiability for the parameter K_{t5} is weak.

❐ **Model appraisal**

The decrease of loss function compared to model $\mathcal{M}_{142}$ is not statistically significant. This yields the following: the model $\mathcal{M}_{142}$ is still not improved. The lower bound of the variance for the parameter K_{ti} are also increased compared to model $\mathcal{M}_{142}$. This means that the identifiability of the parameters has decreased. Consequently, the model $\mathcal{M}_{142}$ is preferred compared to model $\mathcal{M}_{143}$.

4.3.11 Cross validation

➔ **Cross validation using model $\mathcal{M}_{142}$**

We are going to investigate whether the estimated parameters are reproducible. Up to now, the most promising model is $\mathcal{M}_{142}$ and we have estimated the unknown parameters by using measurement sequence 1. The parameter estimation is done also by using the sequence 2.

❐ **Identification**

The estimates based on the sequences 1 and 2 are presented in table 4.8. The identification is only made by using the Extended Kalman Filter for prediction.

❐ **Model analysis**

The parameters K_{bi} seem to be reproducible, but the parameters K_t and B_{toffi} differ quite a lot between sequence 1 and 2. As we can see, the parameter K_t is increased three times for measurement sequence 2.

❐ **Model appraisal**

The submodel describing the flow beside the strip is still promising, but the parameter Kt is very sensitive to measured data. This may be a result of the way we model the flow besides the strip, since this flow is identified by using two parameters. These are dependent of each other, see equation (4.10), and may result in poor identifiability.

Table 4.8 Estimation results, model $\mathcal{M}_{142}$, based on sequences 1 and 2.

	Stochastic	*Stochastic*
Sequence	No. 1	No. 2
K_{b1}	1.2379	0.8454
K_{b2}	0.1448	0.1212
K_{b3}	0.1081	0.1367
K_{b4}	0.1913	0.1690
K_{b5}	0.2898	0.2487
B_{toff2}	7.0000	7.0000
B_{toff3}	2.8808	3.7151
B_{toff4}	3.3691	4.1570
B_{toff5}	5.3380	4.6984
K_t	0.6846	1.8610

➔ **Cross validation using model $\mathcal{M}_{141}$**

The model $\mathcal{M}_{142}$ gives poor reproducible for the parameter K_t. One way to proceed the investigation is to perform the parameter estimation again by letting K_t be a constant. This means that model $\mathcal{M}_{141}$ is used to test whether the parameters is reproducible. Let $K_t \equiv 0.7$, which is a compromise based on identification from other measured sequences not presented here.

❐ **Identification**

The results of the identification are presented in table 4.9 for both sequence 1 and sequence 2.

❐ **Model analysis**

By letting K_t be a constant, even the parameters B_{toffi} are considered reproducible. These parameters are almost the same of the sequence 1 and 2. A ballistic simulation is presented in Fig 4.8, where we can see that the simulated conductivity in the tanks follows the dynamics of the process.

❐ **Model appraisal**

Keeping in mind the purpose of the model, we need a model which can be used to estimate unknown parameters on-line. So far, the model $\mathcal{M}_{141}$ is the most promising for the purpose of the model. However, model $\mathcal{M}_{142}$ is a better one due to the significant decrease in the loss function.

Table 4.9 Estimation results, model $\mathcal{M}_{141}$, based on sequences 1 and 2.

	Stochastic	*Stochastic*
Sequence:	No. 1	No. 2
K_{b1}	1.2402	0.8431
K_{b2}	0.1448	0.1212
K_{b3}	0.1081	0.1063
K_{b4}	0.1907	0.1400
K_{b5}	0.2889	0.2832
B_{toff2}	7.0000	7.0000
B_{toff3}	2.9020	2.8871
B_{toff4}	3.3871	3.3919
B_{toff5}	5.3150	5.6028
K_t	0.7000	0.7000

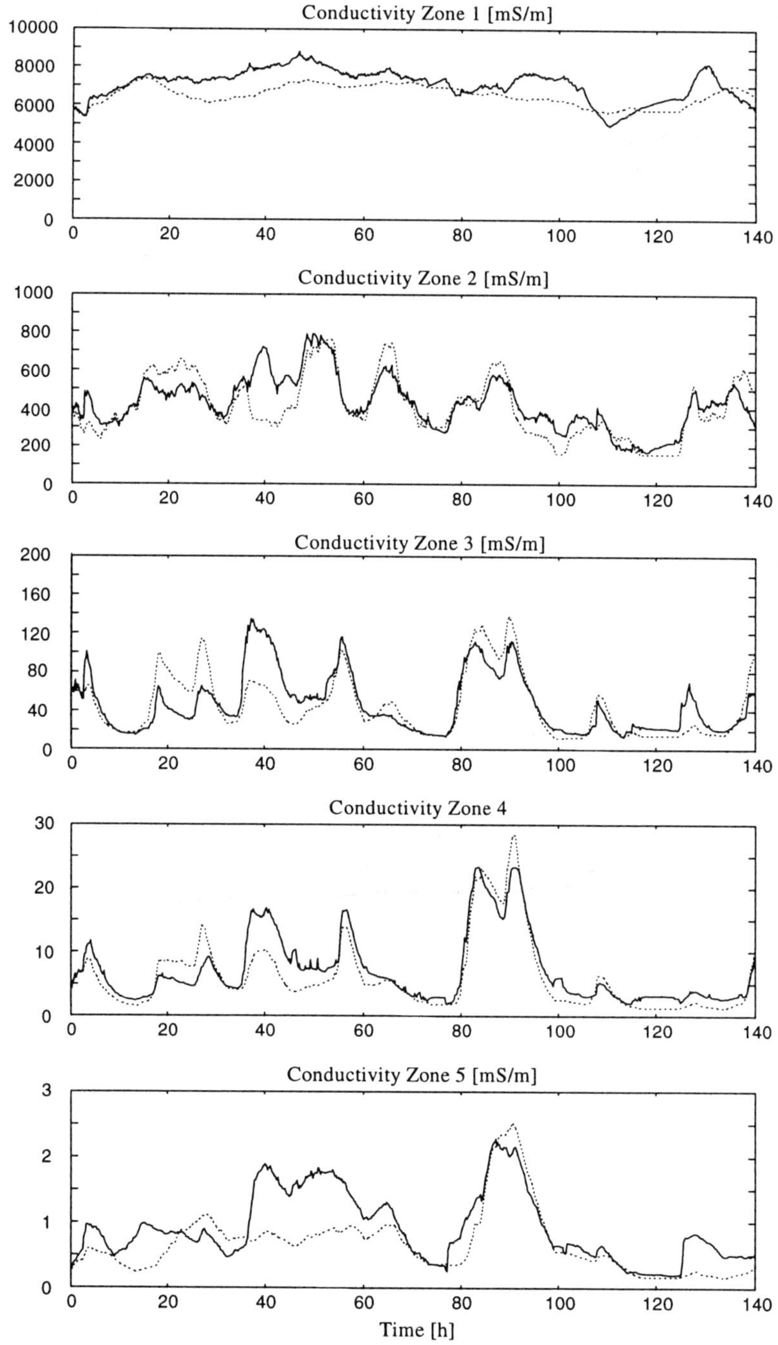

Fig. 4.8 Deterministic simulation of model $\mathcal{M}_{141}$, *dotted* - model output.

4.4 Modelling aspects

From equation (4.9) follows that the dependence of strip thickness is modelled as shown in Fig. 4.9. This figure shows a "knee" at the strip thickness B_{toffi}. At this value, the influence of the strip thickness starts. Consequently, when the strip thickness is below B_{toffi} there is no modelled flow beside the steel strip. For a strip thickness above B_{toffi} the dependence between the modelled flow beside the strip and strip thickness is linear.

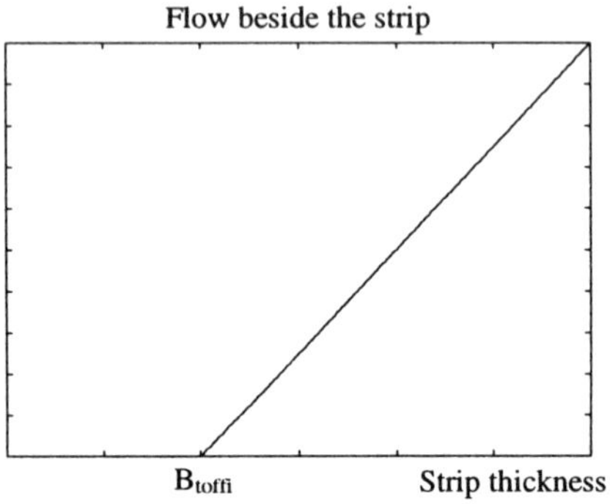

Fig. 4.9 Modelled dependence on strip thickness.

Since the flow via the strip is an uncertain part of the model, other possible model structures have to be investigated. It seems natural to suppose that the flow modelled by equation (4.9) is an approximation of quadratic function, see also Fig. 4.10.

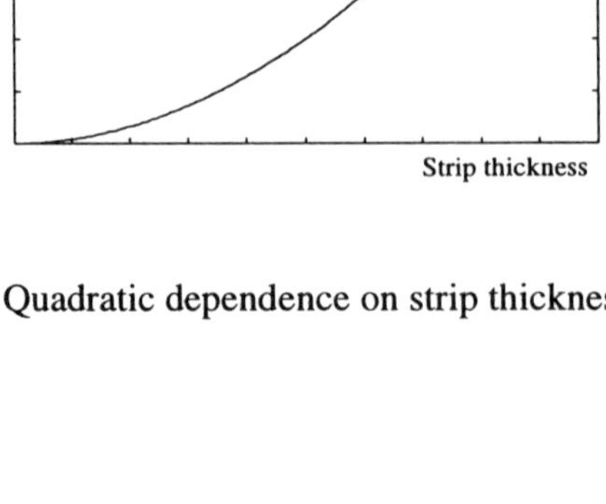

Fig. 4.10 Quadratic dependence on strip thickness.

For a quadratic relationship, the identification with experimental data shows that it is still necessary to have an offset parameter to model the flow beside the steel strip. This means that a quadratic relationship had to be modelled as presented in Fig. 4.11.

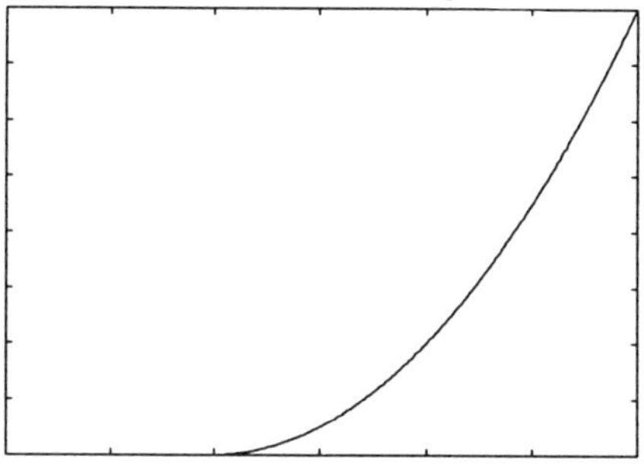

Fig. 4.11 Quadratic dependence on thickness with offset.

A dependence on the strip thickness as shown in Fig. 4.11 gives no better result than the influence of the strip thickness as shown in Fig. 4.9. Consequently, the dependence on strip thickness is chosen as Fig.4.9, because this gives the simplest model of the flow beside the steel strip.

4.5 Alternative tank models

In this section we discuss alternative models of the rinsing process. It is an attempt to expand the structure of the model discussed in Section 4.3. This is done by incorporating incomplete mixing in the rinse tanks or incomplete rinsing of the steel strip.

❒ **Mixing with a "tank in tank" model**

During the rinsing of the steel strip, rinse liquid drops back into the rinse tank and lands on top of the contents of the rinse tank. When the rinse tank is drained off, the surface layer will flow from the rinse tank. If the content of the rinse tank is not completely mixed, a liquid with higher concentration than the other part of the rinse tank will flow from the rinse tank. This should also mean that the concentration of the flow from the rinse tank is dependent on the magnitude of the main flow. Because when the main flow is large, a deeper part of the contents of the rinse tank is involved with the main flow.

Incomplete mixing is modelled by a “tank in tank” model, (Leondes, 1979). This means that the rinse tank is divided into two parts: one upper part which communicates with the main flow and the circulating rinse flow, and a lower part which communicates with the upper part and the circulating rinse flow. In Fig.4.12 the principle of “tank in tank” model is illustrated for rinse zone 3.

The upper part of the rinse tank is represented by $tank_u$ and the lower part is represented by $tank_b$. Concentration and volume of $tank_u$ are represented by C_{u3} and $(1-\gamma)V_3$ and those of $tank_b$ by C_{b3} and γV_3. The circulated flow is represented by F_s. The concentration of the flow dropping back into the rinse tank is represented by C_{b3}.

For the “tank in tank” model we apply the principle of mass balance. The volume of the liquid on the steel strip is very small compared to the circulating flow, $Fs \approx 100$ [m^3/h]. Approximate estimation shows that the time constant of the mixing dynamic on the steel strip is less than 0.1 [s]. Consequently, it is possible to approximate the mixing process on the steel strip as a stiff system.

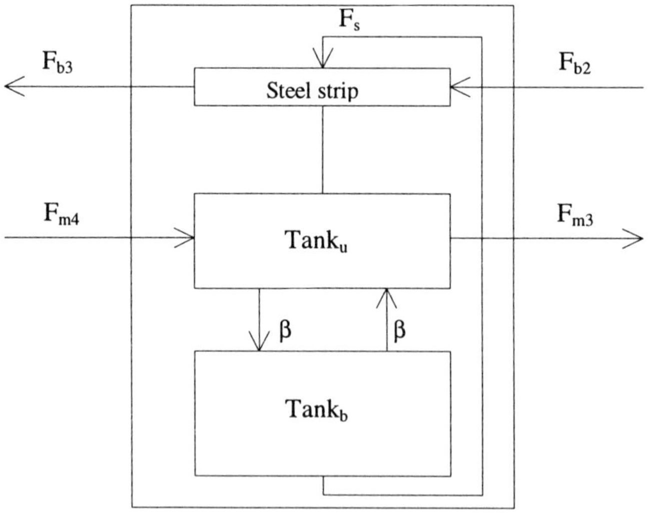

Fig. 4.12 Model of "tank in tank".

Equation (4.12) describes the mixing on the steel strip, equation (4.13) describes the mixing in the upper part of the rinse tank and equation (4.14) describes the mixing in the lower part of the rinse tank.

$$0 = F_{b2} \cdot C_{b2} - F_{b3} \cdot C_{b3} + F_s \cdot C_{b3} - F_s \cdot C_{b3} \tag{4.12}$$

$$(1-\gamma)V_3 \frac{dC_{u3}}{dt} = F_s \cdot C_{b3} - F_{m3} \cdot C_{u3} + F_{m4} \cdot C_{u4} - \beta \cdot C_{u3} + \beta \cdot C_{u3} - F_s \cdot C_{u3} \tag{4.13}$$

$$\gamma \cdot V_3 \frac{dC_{b3}}{dt} = F_s \cdot C_{u3} - \beta \cdot C_{b3} + \beta \cdot C_{u3} - F_s \cdot C_{b3} \tag{4.14}$$

From equation (4.12), C_{b3} is evaluated and put into equation (4.13). Equations (4.13) and (4.14) contain two new unknown constants β and γ. Therefore, with a "tank in tank" model two extra parameters and an extra state have been added to the model of every rinse tank. It may be possible to let β and γ be equal for all the rinse tanks. We have tried to use a "tank in tank" model, but identification of unknown parameters is very time consuming because the model will consist of ten states. It is also difficult to achieve reliable values of β and γ; this model is therefore abandoned.

❐ **Incomplete rinsing of the steel strip**

An attempt was made to model incomplete rinsing of the steel strip. This means that there is a residue concentration on the steel strip from the preceding rinse tank. For example, the concentration on the steel strip after rinse zone 3 is modelled by the following heuristic equation:

$$C_{b3}=(1-S_f)\cdot C_3+S_f\cdot C_2 \tag{4.15}$$

In equation (4.15), S_f is a parameter which gives a relation between the concentration on the steel strip originating from rinse tank 2 and rinse tank 3. Identification with experimental data gives very small and not reproducible values of S_f. This attempt to model incomplete rinsing is therefore abandoned. It should also be mentioned that we have tried to make S_f in equation (4.15) dependent on the strip velocity so the parameter becomes small when the strip velocity is small. This effort was abandoned because the estimation for such a parameter become very small.

4.6 Summary

In this chapter, we start from a basic model of the rinsing process based on a priori knowledge of the process. Since the basic model is not complete, it is developed further by using measured data from the process and results in a grey box model. Obviously, the final model is achieved from the information carried by the measured data. Therefore, the grey box modelling procedure is very much dependent on the data achieved from an experiment with the process.

The resulting model is also a dependent on the model builder, who has to interpret the information from the model analysis. It is a matter of both science and art how to expand a model to produce a better one. The outcome from the grey box modelling procedure also depends on the purpose of the model. It is not always preferable to use the "best" model for an industrial application. In our case, the model is going to be used for on-line estimation of unknown parameters for control and supervision purposes. Therefore, it is necessary to use a model giving reproducible estimates. The result of the grey box modelling procedure is illustrated by the following tree structure:

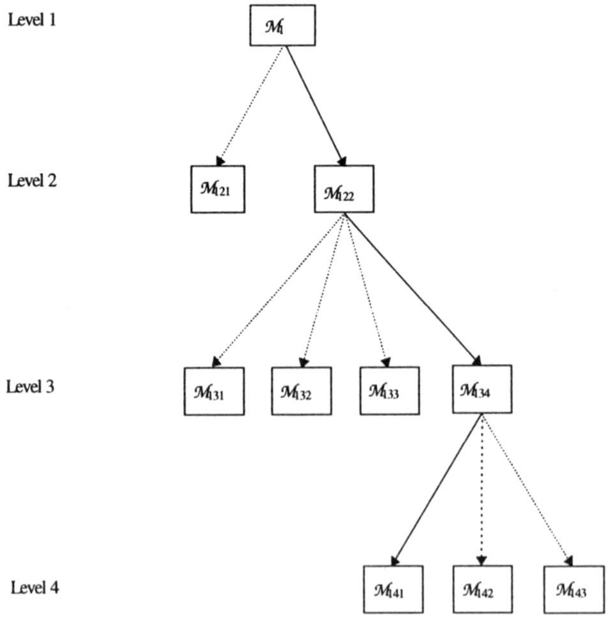

Fig. 4.13 Tree structure from grey box modelling of the rinsing process.

5 REAL-TIME ESTIMATION

5.1 Introduction

Recursive identification methods are necessary to solve the real-time estimation problem. The subject is treated in detail by, for example Andersson and Moore (1979) or Ljung and Söderström (1983). Theoretical and practical aspects of Kalman filters are presented by Grewal and Andrews (1993). Unknown parameters of a process model have to be estimated on-line, because they will be used when we supervise the efficiency of the parts of the process subject to wear and due to parameter variations. The estimates will also be used for process control.

In Section 5.2, the data acquisition is treated. We are interested in what way replacement of worn parts influence the efficiency of the process. Therefore, data are collected before and after a replacement of worn parts. The measurements are done when the process is affected with the controller normally used, since this is the circumstance the on-line estimation is going to be applied.

In Section 5.3, the identification is done off-line to achieve estimates for comparison. In Section 5.4, the unknown parameters of the model are estimated by an on-line estimation method. Extended Kalman Filter is used to estimate both the state variables and parameters of the process. The filter is simulated with the experimental data collected during conventional running of the process.

5.2 Data acquisition

The purpose of collecting measurements from the process is to obtain data which can be used for estimation of unknown parameters. Our interest in data acquisition is focused on the question of in what way the replacement of parts subject to wear influences the behaviour of the process. Furthermore, it is of interest to investigate whether the efficiency of the process deteriorates during the period between two planned stops for maintenance.

For the rinsing process, the flow of clean water into tank 5 is controlled by a simple feedforward controller. The flow of clean water is computed from equation (5.1), where K_f=17.5 10^{-6} is a calibration constant, B_v [m/min.] is strip velocity, and B_b [mm] is strip width.

$$F_{c1}=K_f \cdot B_v \cdot B_b \qquad (5.1)$$

The flow of clean water into tank 2 is controlled by an on/off controller. When the conductivity value in tank 2 is above 500 [mS/m] or the liquid level in tank 2 is below 0.5 [m] an extra flow of 1.0 [m^3/h] is fed direct into tank 2. Otherwise, this flow is zero.

When logging the process inputs and outputs during normal operation of the process, it is seen that the outputs vary considerably. The outputs seem to be informative enough to be used to estimate the unknown model parameters, see Fig. 5.2. So, the flow of clean water into tank 5 and tank 2 is not perturbed by any special input sequence during the experiment outside normal operation

Considering the discussion above, one measurement sequence is collected before a planned stop, and the other measurement sequence is collected after the replacement of rolls. Furthermore, a third measurement sequence is collected directly after the second measurement sequence to study the effect of any wear of the squeezer rolls between two planned stops for maintenance. The three measurement sequences are denoted; sequence 1, sequence 2 and sequence 3, see Fig. 5.1.

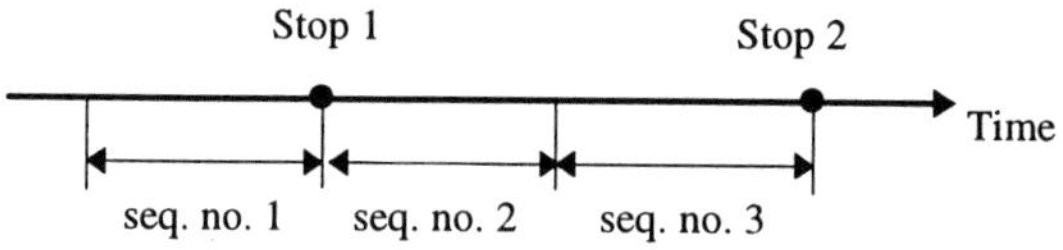

Fig. 5.1 Experiment sequences.

The number of samples may not be too short, since we want to investigate the time variant behaviour of the on-line estimates. Sequence 1 consists of 570 sample points, which corresponds to 4.7 days. After collecting sequence 1, a planned stop for maintenance occurred and the pairs of squeezer rolls 1, 2, 4 and 6 were replaced by renovated squeezer rolls. Sequence.2 consists of 800 sample points, which corresponds to 6.7 days. Sequence 3 consists of 700 sample points, which corresponds to 5.8 days. Note that this sequence was collected directly after sequence no. 2 without any planned stop for replacement of squeezer rolls.

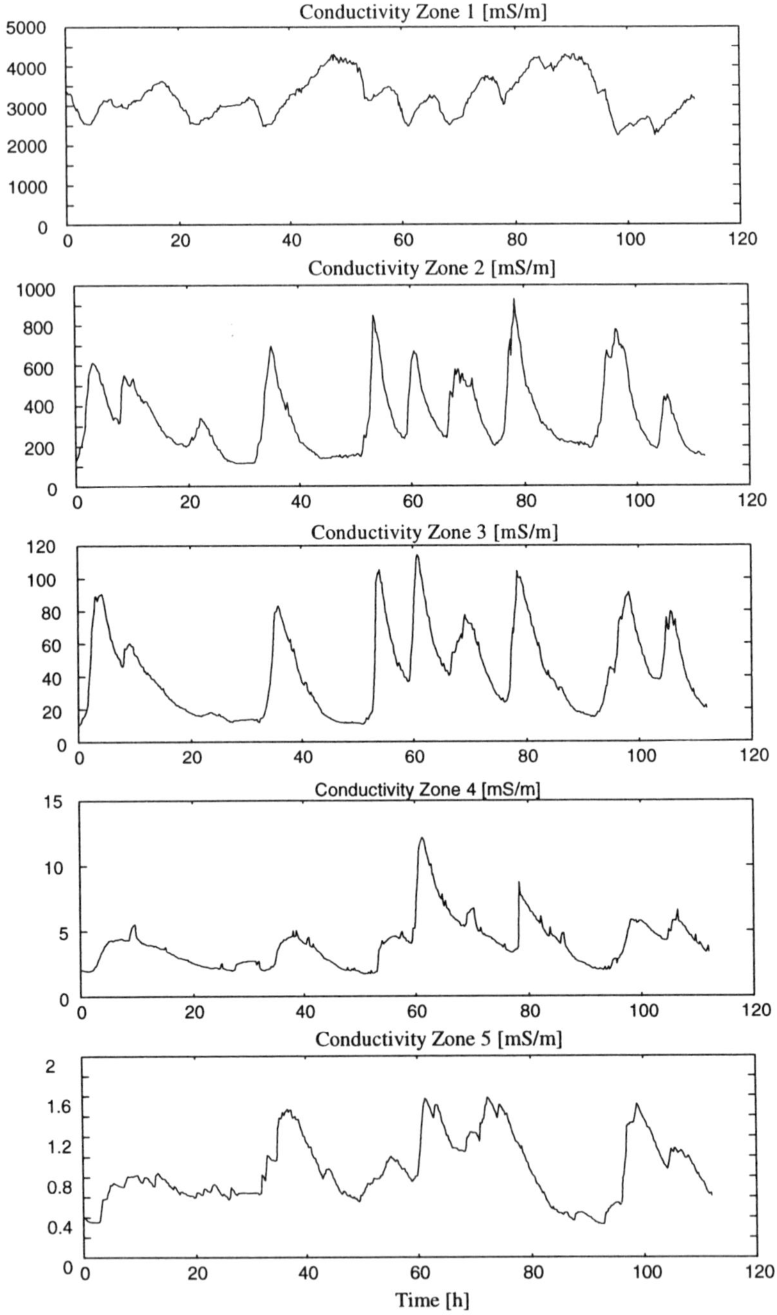

Fig. 5.2 Conductivity measurements, sequence 1.

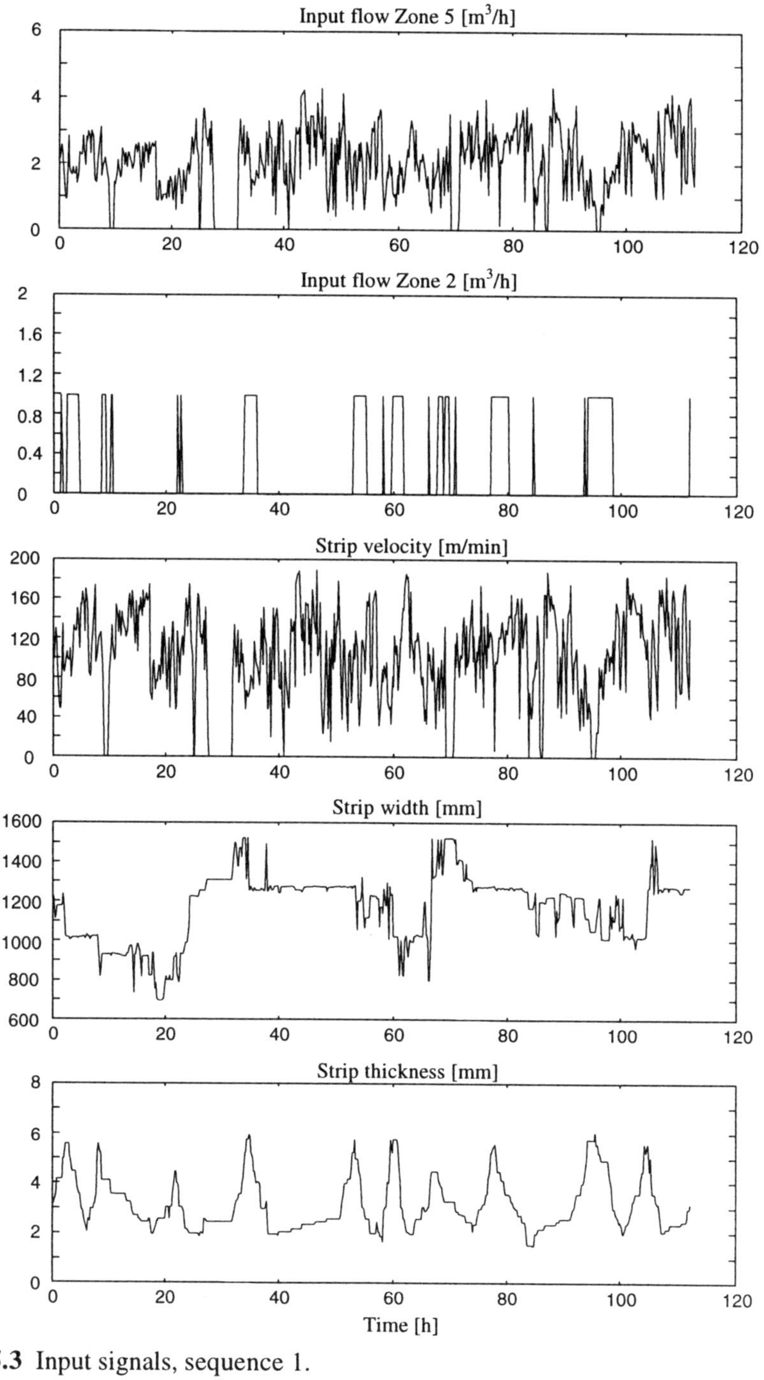

Fig. 5.3 Input signals, sequence 1.

❑ **Comments on sequence 1.**

The conductivity in rinse tank 2 varies considerably; the peaks of the amplitude coincide with those of the strip thickness, see Fig. 5.2 and Fig. 5.3. These variations also appear in rinse tank 3, 4 and 5. Note that the conductivity in tank 1 varies out of phase compared with the strip thickness. When a thick steel strip is rinsed, the conductivity decreases, while, when a thin strip is rinsed the conductivity is increased. This is explained by the efficiency of the squeezer rolls 2. When a thick strip passes these squeezer rolls, we have a large flow beside the strip. The flow via the strip from tank 1 will cause the liquid level in tank 1 to decrease. To compensate this, rinse liquid with a lower concentration is pumped from tank 2 to tank 1. Consequently, the conductivity in tank 1 is decreased.

The strip thickness has a typical variation between thin and thick steel strip, see Fig. 5.3. The strip thickness varies between 2 - 6 [mm]. The strip width also has a typical variation between narrow and broad steel strip. The strip width varies between 600 -1600 [mm].

The strip velocity varies frequently. This is typical of the way the process is operated and depends mainly on the fact that it takes time to weld the ends of two coils at the beginning of the process. It also takes time to split the steel strip into coils again at the end from the process. The entry and delivery loop cars cannot absorb all the delay. The strip velocity varies between 0 - 200 [m/min.]. Note that after twenty-eight hours, there was a stop of four hours. This was not a planned stop and none of the squeezer rolls were replaced.

The flow of clean water into tank 5 has a variation typical of the way the process operates, this means that the flow is computed as the product of strip velocity and strip width, see equation (5.1). The flow of clean water varies between 1.0 - 4.0 [m^3/h].

The flow of clean water into tank 2 is turned on when the conductivity in tank 2 is above 500 [mS/m]. For sequence no.1, this happens relatively often, depending on the high conductivity amplitude. Some of the shorter admissions arise when the level of rinse liquid in tank 2 is below 0.5 [m].

❑ **Comments on sequence 2.**

This sequence is not presented, but some comments are given. Note that a planned stop for maintenance has taken place between sequence 1 and sequence 2. The squeezer rolls 1, 2, 4 and 6 have been replaced by other, renovated rolls during this stop.

The conductivity in tank 2 no longer has any high amplitudes coinciding with the strip thickness as sequence 1 has. The conductivity in rinse tank 3 has variation coinciding with that of the strip thickness, but the amplitudes are lower compared to sequence 1. The degree of conductivity in tank 4 and 5 also have variations. The conductivity in tank 1 is different from that in sequence 1.

The strip thickness and strip width have typical variations similar to sequence 1. The strip velocity varies considerably and in the same way as in sequence 1.

The flow of clean water into tank 5 is also similar to that of sequence 1. However, the flow of clean water into tank 2 is different compared to sequence 1. This is because the conductivity in tank 2 has not exceeded the limit for the flow to be turned on during sequence 2. However, some brief admission of water has taken place because the level of rinse liquid in tank 2 was too low in the tank.

❑ **Comments on sequence 3.**

This sequence is not shown, but some comments are given. This sequence follows directly after sequence 2. Only a brief stop has taken place to inspect the measurement devices. No replacement of squeezer rolls has taken place between sequence 2 and sequence 3.

The conductivity rates in the rinse tanks show comparatively similar behaviour to those of sequence 2. However, the conductivity amplitude in tank 4 has increased. Also, the mean level of conductivity in tank 5 has increased. The other inputs have roughly similar behaviours to those in sequence 2.

5.3 Off-line identification

The off-line identification based on model $\mathcal{M}_{141}$ and the identification is done by optimising the likelihood function, see Chapter 4. The results of the estimation are shown in Table 5.1. Notice, the squeezer rolls 1 and 5 are represented by only one parameter and the squeezer rolls 2, 3 and 4 are represented by two parameters. The pair of squeezer rolls 5 is represented by one parameter, since we found it is difficult to perform a robust on-line estimation of two parameters. A replacement of squeezer rolls are indicated with the symbol: ☞

Table 5.1 Off-line identification.

Squeezer rolls	Parameter.	*Sequence No.1*	*Changed rolls*	*Sequence No.2*	*Sequence No.3*
No. 1	K_{b1}	0.82	☞	0.60	0.65
No. 2	K_{b2}	0.20	☞	0.22	0.22
No. 2	B_{toff2}	2.80	☞	5.08	4.87
No. 3	K_{b3}	0.42	-	0.32	0.36
No. 3	B_{toff3}	4.06	-	3.43	3.44
No. 4	K_{b4}	0.34	☞	0.51	0.60
No. 4	B_{toff4}	4.86	☞	4.42	4.42
No. 5	K_{b5}	0.86	-	1.24	1.62

❐ **Comments on the pair of squeezer rolls 1.**

These squeezer rolls were replaced during the planned stop for maintenance. It is seen from Table 5.1, that the new squeezer rolls are more efficient than the old rolls. This means that the liquid film on the strip became thinner after the change of squeezer rolls. In a comparison between the estimates from the two latest sequences, there is a possible decrease in efficiency from sequence 2 to sequence 3. This can indicate that these squeezer rolls are becoming worn while the process is running.

❐ **Comments on the pair of squeezer rolls 2.**

These squeezer rolls were replaced during the planned stop. It is seen from Table 5.1, that B_{toff2} is about three before the change and about five after the change. This means that the squeezer rolls used before the change were harder than the new squeezer rolls. Consequently, the flow beside the strip was reduced after the change of rolls. It is also seen from Table 5.1 that the parameter B_{toff2} is reduced between sequence 2 to sequence 3. This means that the rubber became harder when the process was running, causing the flow beside the strip to increase. It is known from experience that the rubber of the rolls becomes harder when they are used.

Concerning the parameter K_{b2}, the thickness of the liquid at the top and bottom of the strip, this parameter has about the same value before and after the exchange of squeezer rolls. The exchange of squeezer rolls has not influenced this parameter considerably. This means that the surfaces of the old rolls were not too worn.

❐ **Comments on the pair of squeezer rolls 3.**

These squeezer rolls were not replaced during the planned stop for maintenance. It is seen from Table 5.1, that the parameter value B_{toff3} has decreased between sequence 1 and sequence 2. This means from a physical point of view, that the squeezer rolls became harder when the process was running and the flow beside the strip has increased.

❐ **Comments on the pair of squeezer rolls 4.**

These squeezer rolls were replaced during the planned stop for maintenance. The estimate of the parameter B_{toff4} is less for the new squeezer rolls; this means that the new rolls are harder than the old ones. K_{b4} is considerably larger for the new squeezer rolls, which means that the flow via the top and bottom of the strip is larger for the new rolls. The estimates show that the new rolls are less efficient than the older squeezer rolls.

❐ **Comments on the pair of squeezer rolls 5.**

These squeezer rolls were not replaced during the planned stop for maintenance. The parameter K_{b5} is increased between all three sequences. The increase is relatively large. This means that this pair of squeezer rolls has been worn and the flow via the strip has increased.

5.4 On-line estimation

On-line estimation is based on the Extended Kalman Filter, (EKF). The filter consists of a linearization of the state equations around the estimated state at every discrete sample instant. It is an extension of the linear Kalman filter for linear systems, (Jazwinsky, 1970).

The rinsing process is described by the following equations:

$$x(k+1) = F\left[u(k),\ s(k),\ \theta(k)\right] + v(k) \tag{5.2}$$

$$y(k) = G\left[x(k)\right] + w(k) \tag{5.3}$$

where k is the sample index, v(k) and w(k) are assumed to be independent discrete white noise with zero mean value and covariance matrices R_1 and R_2.

The vector θ(k), consists of the unknown parameters to be estimated:

$$\theta(k) = [K_{b1}\ K_{b2}\ B_{toff2}\ K_{b3}\ B_{toff3}\ K_{b4}\ B_{toff4}\ K_{b5}\ ]^T \tag{5.4}$$

Generally, θ(k) is modelled by the following difference equation, (La Cava, 1989):

$$\theta(k+1) = \theta(k) + v_\theta(k) \tag{5.5}$$

where θ(0) is a Gaussian initial value with mean value $\hat{\theta}(0)$ and covariance matrix $P_\theta(0)$. The discrete white noise is represented by $v_\theta(k)$, with zero mean and covariance matrix R_θ.

Introduce an augmented state vector:

$$\xi(k) = [x(k)^T\ \theta(k)^T]^T \tag{5.6}$$

The state equations of the augmented system are described by the following system:

$$\xi(k+1) = \overline{F}\left[\xi(k),\ u(k),\ s(k)\right] + v_\xi(k) \tag{5.7}$$

$$y(k) = \overline{G}\left[\xi(k)\right] + w(k) \tag{5.8}$$

where the process noise of the augmented state vector is $v_\xi(k)$ with covariance matrix R_ξ,

$$v_\xi(k) = [v(k)^T \; v_\theta(k)^T]^T \qquad (5.9)$$

$$R_\xi = \begin{bmatrix} R_1 & 0 \\ 0 & R_\theta \end{bmatrix} \qquad (5.10)$$

Application of an EKF on the augmented system, estimate both the process states and the unknown parameters. The same algorithm is used as presented in Section 2.3. At start-up of the Extended Kalman Filter, the process states are initialised at likely values. Note that the process states are scaled.

$$x(0) = [0.5\ 0.5\ 0.5\ 0.5\ 0.5]^T.$$

The vector of the unknown parameters is initialised at likely values:

$$\theta(0) = [0.5\ \ 0.5\ \ 4.0\ \ 0.5\ \ 4.0\ \ 0.5\ \ 4.0\ \ 0.5]^T.$$

These values are chosen in order of size as obtained during off-line parameter estimation. The initial value of the covariance matrix of the estimation error is chosen at:

$$P(0) = \mathrm{diag}[\ 0.1\ \ 0.1\ \ 0.1\ \ 0.1\ \ 0.1\ \ 0.1\ \ 0.01\ \ 0.1\ \ 0.01\ \ 0.1\ \ 0.01\ \ 0.1\ \ 0.1\ ].$$

The first five values in P(0) are assigned to the states of the process and the eight last values of P(0) are assigned to the parameter vector θ. The values are chosen by repeated simulation so $\theta(k)$ will have an acceptable settling time without unnecessary overshoots. The choice of P(0) is not critical. The covariance matrices of R_1 and R_2 are set to the same values as were used during off-line parameter estimation, while R_θ is chosen by repeated simulations until acceptable results have been achieved for the parameter vector $\theta(k)$. Note that the elements of R_θ are chosen at smaller values than the values of R_1. This means, that the process disturbances will be modelled mainly by v(k) and less by $v_\theta(k)$. From Section 4.3 R_1 and R_2 are chosen as:

$$R_1 = \mathrm{diag}[1\cdot 10^{-3}\ \ 1\cdot 10^{-3}\ \ 1\cdot 10^{-3}\ \ 1\cdot 10^{-3}\ \ 1\cdot 10^{-3}]$$

$$R_2 = \mathrm{diag}[1\cdot 10^{-4}\ \ 1\cdot 10^{-4}\ \ 1\cdot 10^{-4}\ \ 1\cdot 10^{-4}\ \ 1\cdot 10^{-4}]$$

From the simulation of the Extended Kalman Filter, suitable values of R_θ are:

$$R_\theta = \text{diag}[1\cdot10^{-6}\ \ 1\cdot10^{-6}\ \ 1\cdot10^{-6}\ \ 1\cdot10^{-6}\ \ 1\cdot10^{-6}\ \ 1\cdot10^{-6}\ \ 1\cdot10^{-6}\ \ 1\cdot10^{-6}]$$

Estimation of the unknown parameter vector based on sequence 1 is shown in Fig. 5.4 and Fig. 5.5. Sequence 2 and sequence 3, collected after the planned stop for maintenance are put together into a joint sequence. This sequence consists of 1500 samples or 12.5 days. With a long measurement sequence, it is possible to investigate whether the squeezer rolls wear out while the process is running. In Fig. 5.6 and Fig. 5.7, the estimated parameter vector $\theta(k)$ is presented for the long sequence.

- **Comments on the pair of squeezer rolls 1.**

  This pair of squeezer rolls was replaced during the planned stop. By comparing the results from sequence 1 and the merged sequence, it is obvious that K_{b1} is smaller for the new pair of squeezer rolls. The result agrees with the result from the off-line estimation of K_{b1}.

- **Comments on the pair of squeezer rolls 2.**

  This pair of squeezer rolls was replaced during the planned stop. A comparison between the results from sequence 1 and those of the merged sequence shows that a change of this pair of squeezer rolls has a large influence on the parameter B_{toff2}. The values from the on-line estimation are similar to those obtained from off-line estimation. It is not possible to detect any wear out of the squeezer rolls for the merged sequence.

- **Comments on the pair of squeezer rolls 3.**

  This pair of squeezer rolls was not replaced during the planned stop. The estimation of the unknown parameters for this pair agrees with the results from the off-line estimation. The decrease in B_{toff3} between sequence 1 and sequence 2 is obtained from both off-line and on-line parameter estimation.

- **Comments on the pair of squeezer rolls 4.**

  This pair of squeezer rolls was replaced during the planned stop. The results of estimation of the parameters is about the same as from off-line parameter estimation. It should be pointed out that the new squeezer rolls are worse than the old ones. For the merged sequence we can see that the value of K_{b4} increases during the second half of the merged sequence; this is also detected from the off-line estimation. The increase is considerably larger than the corresponding standard deviation of the estimation error. This means that this pair of squeezer rolls deteriorated when the process was running. Note also that the on-line estimation shows that the wear is rather large after 200 hours and not after 160 hours, which the off-line parameter estimation is based on.

❒ **Comments on the pair of squeezer rolls 5.**

This pair of squeezer rolls was not replaced during the planned stop. From the merged sequence it is shown that this pair of squeezer rolls deteriorated when the process was running.

To sum up: the results from the on-line estimation show that the values of the parameter vector converge to similar values to those achieved from the off-line estimation. The settling time depends on the initial values $P_\theta(0)$ and R_θ. It can be pointed out that the change in the efficiency of the squeezer rolls can be detected faster if R_θ is larger. However, larger values of R_θ mean that the process disturbances will influence the estimates more, and there will be more noise in the estimates.

It should also be pointed out that the parameters according to the flow at the side of the strip occasionally changes abruptly. This is a consequence of the way the flow beside the strip is modelled. The strip thickness had to exceed the value of the estimated parameter before the flow beside the strip has any influence on the model.

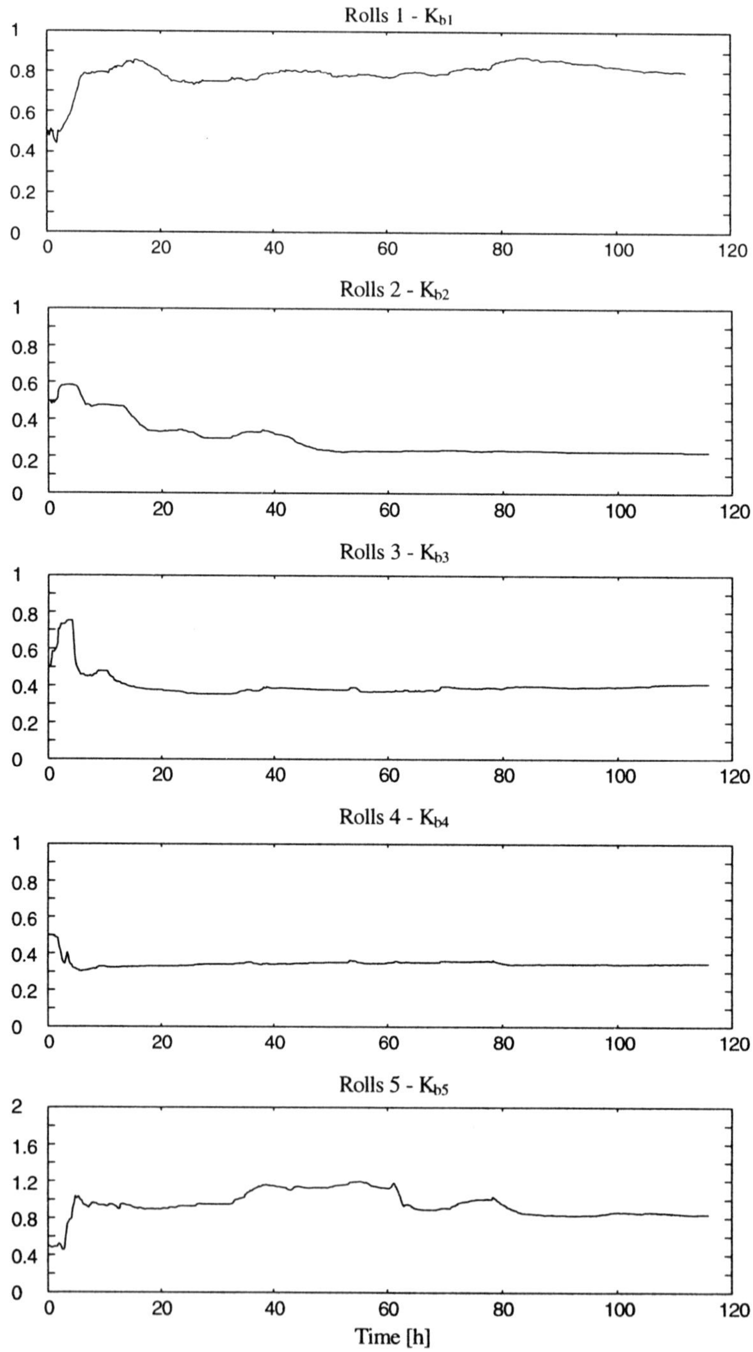

Fig. 5.4 On-line estimation of parameter K_{bi}, sequence 1.

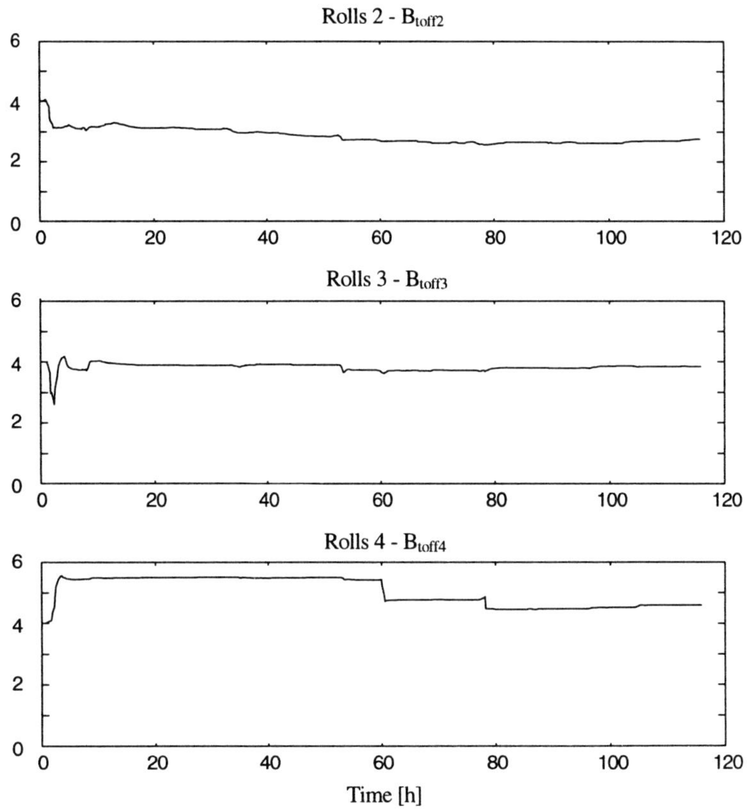

Fig. 5.5 On-line estimation of parameters B_{toffi}, sequence 1.

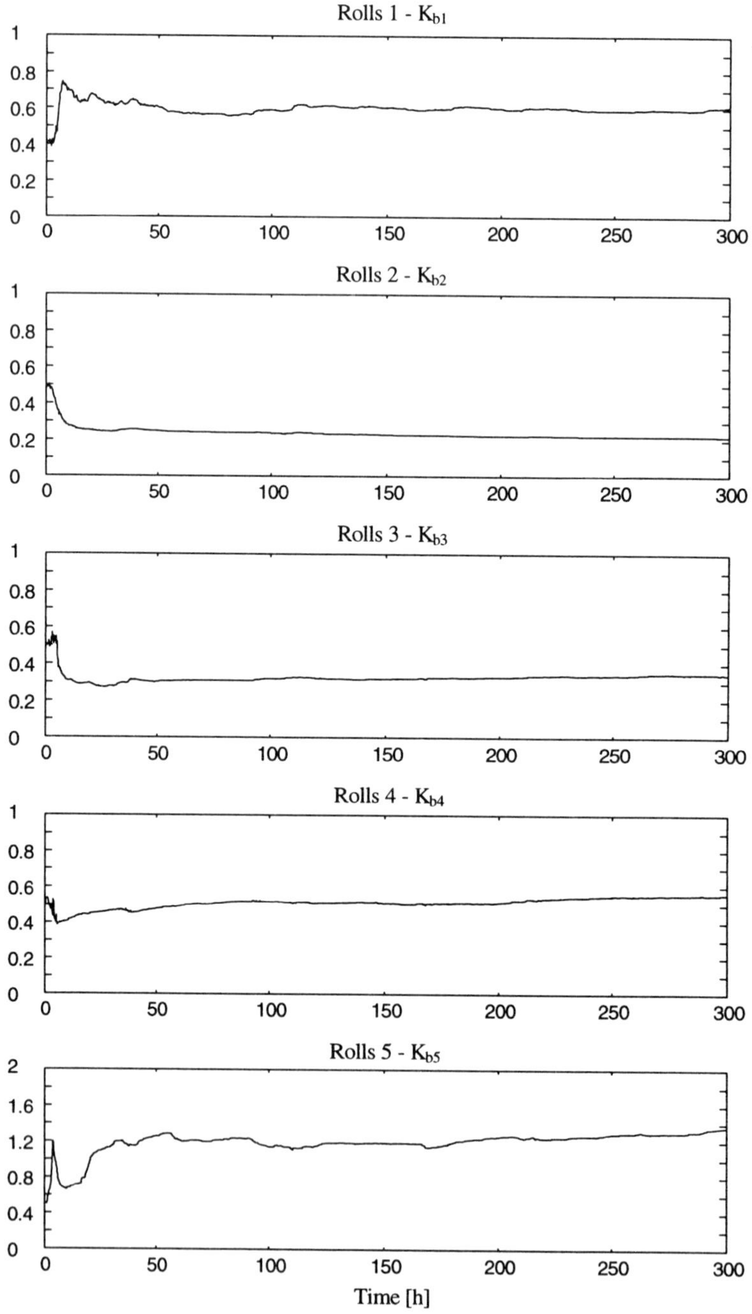

Fig. 5.6 On-line estimation of parameter K_{bi}, sequence 1 + sequence 2.

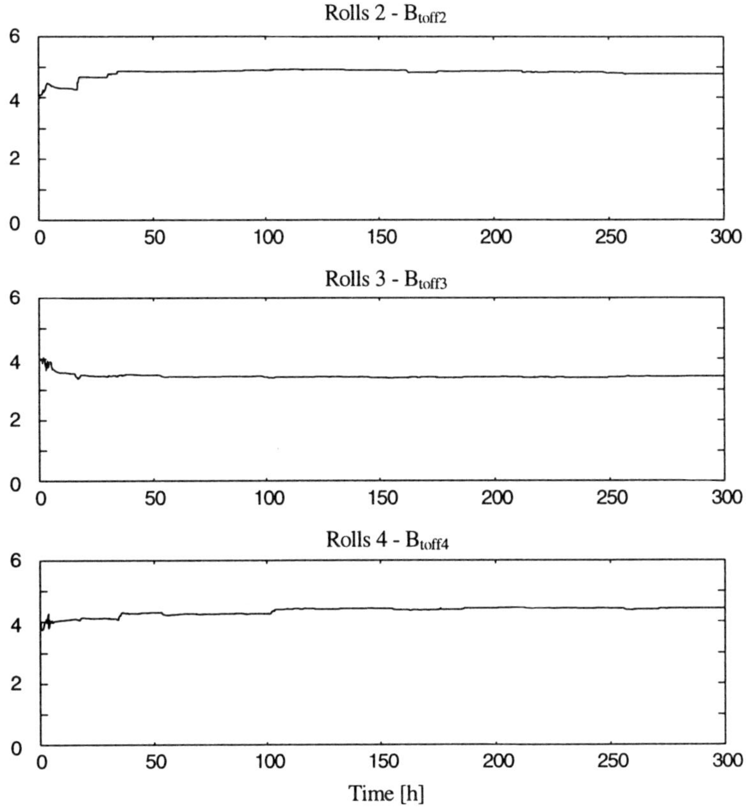

Fig. 5.7 On-line estimation of parameter B_{toffi} , sequence 1 + sequence 2.

5.5 Summary

Prerequisite to cope with wearing parts, is to estimate the wearing on-line. Grey box modelling is based on physical insight and give a basis for on-line estimation. For the rinsing process, it is possible to model the wearing parts explicitly by grey box modelling. With this method we can distinguish the behaviour of the parts subject to wear from other parts of the process.

Off-line identification of the unknown parameters, shows that the replacements affect the efficiency considerably. It is also shown that it is a non-trivial task to decide whether a part is worn or not and also be replaced or not. The estimation of the unknown parameters is also made on-line by the Extended Kalman Filter. The state vector of the process is augmented by the vector of the unknown parameters.

From the simulation with the same measurement data, used during off-line parameter estimation, it is shown that the Extended Kalman Filter estimates the unknown parameters well. During off-line parameter estimation, some changes in the efficiency of the squeezer rolls were detected. The Extended Kalman Filter detects these changes also. To sum up, the Extended Kalman Filter can be used to estimate a finite number of unknown parameters in the process model. This offers an opportunity to display an easily interpreted measure of the efficiency of the process to the operators. An adaptive controller can also use values of the parameters estimated on-line.

6 PROCESS CONTROL

6.1 Introduction

The purpose of controlling many processes within the steel industry is to retain important of process variables on specified levels so the final product fulfils given specifications. The control actions to obtain such a result are often associated with some costs. This means that the control engineer has limited resources to design with when constructing and tuning a control system. There is also demand for a consistent production with a low rate of re-processing and waste.

Increasing competition within the steel industry necessitates that the time between the order and the delivery is short and precise. This means that the control system must be able to manage a flexible production with fast changes in the production mix. Consequently, typical of the steel production is the influence of changing production parameters, for example product dimensions and product mix.

It is well known that the economic operating point of a process often lies at the intersection of the product constraints. A successful controller must therefore run the process as close as possible to such constraints without violating them subject to unpredictable disturbances and noise. In addition, the production costs are affected by the condition of the process, maintenance, accessibility and failure. These types of costs are discussed further in Chapter 7.

In this chapter, we demonstrate two different types of controllers. Both types are adaptive, since the unknown parameters of the model are estimated on-line by the Extended Kalman Filter as described in Chapter 5.

The first is a combination of feedforward and feedback control. The controller is presented in Section 6.2 and tested in Section 6.3. This controller is in a way heuristic, since it should be easily implemented. The feedforward part compensates for the variation in the production parameters; strip velocity, strip width, and strip thickness. The feedback part compensates for disturbances acting on the process. Furthermore, the feedback part compensates for the fact that the feedforward controller cannot be made perfect.

The second type is an optimal controller and is based on a more general model predictive control methodology. The controller is presented in Section 6.4 and tested in Section 6.5. The optimisation problem is solved at each sample instance using new information about the production parameters and estimated parameters. The process inputs are computed over a future time interval including all process restrictions.

6.2 Combined feedforward and feedback controller

The dynamics of a non-linear process are often related to exogenous inputs or other variables. A commonly used approach to control such systems is to apply the method of gain scheduling. This method is based on the controller gain being adapted to discrete values depending on variations of the process dynamics and exogenous inputs. The process model is linearized around several operating points, originating from a set of the exogenous inputs. Linear control theory is applied to the set of linearized models and a table of parameters for a control law is computed. When the process is controlled, the exogenous signal affects a pointer to pick up the corresponding parameters from the table and change the parameters of the control law. An introduction to gain scheduling is given by Åström and Wittenmark (1995) and the principle is illustrated in Fig. 6.1.

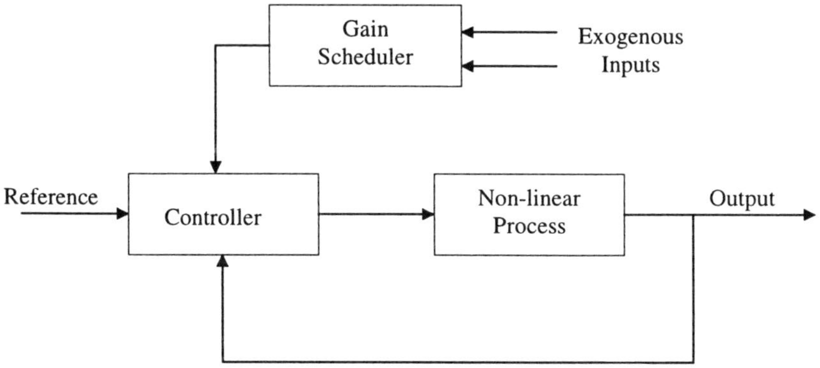

Fig. 6.1 Gain scheduling scheme.

Gain scheduling has been successfully applied to flight control systems and later on for control of industrial processes too. The method has been popular since it is simple to apply and the computational burden is small as compared to other more advanced controllers. Despite its popularity, gain scheduling is still an ad hoc methodology. Theory of global gain scheduling is fragmentary. It can be pointed out that gain scheduling is an example of bridging the gap between theory and practice, but in this situation application is ahead of theory, (Rugh, 1991).

Some guidance in designing a gain scheduling controller is given in the literature, see e.g. Shamma and Athens (1992). It concern the influence on the non-linearity of the process and the influence of exogenous or scheduling variables. The two basic guidelines are:

- The scheduling variable should capture the non-linearity of the process
- The scheduling variable should vary slowly.

A framework for gain scheduling is presented by Rugh (1991). It is based on a non-linear process model and on a general system description:

$$\dot{x}(t) = f\left[x(t), u(t), s(t)\right] \tag{6.1}$$

$$y(t) = g\left[x(t), u(t), s(t)\right] \tag{6.2}$$

where x(t) is an n-dimension state vector, u(t) is a r-dimensional input control vector, y(t) is a p-dimensional output vector and s(t) is a q-dimensional exogenous scheduling variable. It is also assumed that the non-linear functions f and g are continuously differentiable.

The control problem is evaluated by computing a control signal as a function of the states and the scheduling variable:

$$u(t) = \mathcal{f}[x(t), s(t), r_{ref}(t)] \tag{6.3}$$

where $\mathcal{f}$ is also a non-linear continuously differentiable function.

This controller is typically non-linear and is a gain scheduling controller due to the dependence on the exogenous signal s(t) and reference signal r(t). A procedure to design the controller is presented by Lawrence and Rugh (1995). In general terms is described by the following steps:

1. Compute a set of constant operating points of the process model based on a set of constant values of the process variables and exogenous inputs.
2. For the set of operating points, design a set of linear controllers.
3. Compute a gain scheduling controller such that at each operating points, the controller provides a constant control signal, which gives zero error.

4. Analyse non-local performance by creating the behaviour of the gain scheduling controller by simulation.

For a set of constant exogenous inputs s^o and reference values, there is a set of stationary points (x^o, u^o), which are the steady state solutions of the non-linear equations describing the process. The solutions are achieved by requiring the time derivative in equation (6.1) to be zero:

$$0 = f\left[x^{o}, u^{o}, s^{o}\right] \qquad (6.4)$$

By letting a specific necessary number of the states be references, a single unambiguous solution of equation (6.4) is achieved. The steady state solution gives a corresponding value on the control signal u^o. This signal can be seen as a feedforward action, which compensates for changes in the exogenous signal.

In the ideal situation with no unmeasurable disturbances and with a perfect model, the feedforward signal u^o would be enough to control the process. However, for a real plant there are unmeasurable disturbances acting on the process which will cause errors between the reference and controlled variable. Furthermore, the process model is never a perfect copy of the process. When constructing a model, we have to make several simplifications and guesses to achieve a useful model. To take care of these defects, we also introduce a feedback controller. The controller described in Fig. 6.1 can accordingly be divided into two parts; a feedforward and a feedback part, see Fig. 6.2.

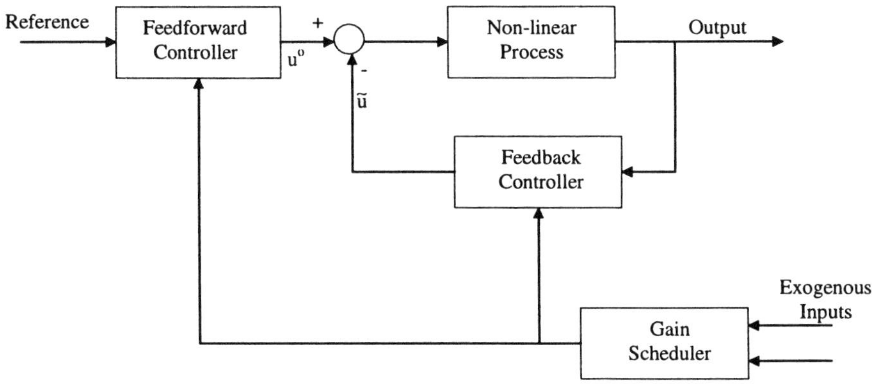

Fig. 6.2 Feedforward and feedback controller.

To simplify and reduce the computational burden, it is more convenient to use a discrete-time equivalent model, (Baumann, 1991). Continuous-time control needs a very fast sampling rate, but for discrete-time control the sampling rate is a design variable and can be set at a suitable value by the control engineer. The controller will be based on the discrete equivalent model:

$$x(k+1) = F\big[x(k), u(k), s(k)\big] \tag{6.5}$$

$$y(k) = G\big[x(k), u(k), s(k)\big] \tag{6.6}$$

where F and G are non-linear differentiable functions. The variables have the same meaning as before, but u(k) is assumed to be constant during the sample intervals.

The stationary points are achieved when x(k+1)=x(k), which is set equal to x^o. The solution to equation 6.7 gives also the feedforward signal u^o , as described in Fig. 6.2. As for the continuous case, we search for the steady state solution for a set of exogenous inputs s^o and reference values by letting:

$$x^o = F\left[x^o, u^o, s^o\right] \tag{6.7}$$

To develop the feedback controller, the non-linear model given by equation (6.5) and (6.6) is linearized, (Kwakernaak & Sivan, 1972). The linearisation is made around the stationary values x^o and u^o. Introduce the states and the control signals as the sum of the stationary values, and the deviations $\tilde{x}(k)$ and $\tilde{u}(k)$.

$$x(k) = x^o + \tilde{x}(k) \tag{6.8}$$

$$u(k) = u^o + \tilde{u}(k) \tag{6.9}$$

Introduce also the matrices Φ and Γ, which are the gradients of the non-linear function F with respect to x(k) and u(k), taken at the stationary points x^o and u^o.

$$\Phi = \left.\frac{\partial F}{\partial x}\right|_{x^o, u^o} \tag{6.10}$$

$$\Gamma = \left.\frac{\partial F}{\partial u}\right|_{x^o, u^o} \tag{6.11}$$

The linearized model is then given by equation (6.12).

$$\tilde{x}(k+1) = \Phi\,\tilde{x}(k) + \Gamma\,\tilde{u}(k) \qquad (6.12)$$

A Linear Quadratic (LQ) controller is applied to the linearized model of the process, and a search is made for a control sequence $\tilde{u}(k)$, k=1,..., N-1, which minimises a quadratic loss function, (Åström and Wittenmark, 1990), which is given by:

$$J = E\left\{\sum_{k=1}^{N-1}\left[\tilde{x}(k)^T Q_1 \tilde{x}(k) + \tilde{u}(k)^T Q_2 \tilde{u}(k)\right] + \tilde{x}(N)^T Q_N \tilde{x}(N)\right\} \qquad (6.13)$$

where N is the number of samples, Q_N, Q_1 and Q_2 are weighting matrices, which define the penalty of the deviation from the reference values. The controller is based on (6.13) and is given by equation (6.14).

$$\tilde{u}(k) = -L(k)\tilde{x}(k) \qquad (6.14)$$

The vector L(k) is computed from equations (6.15) - (6.17):

$$L(k)=[Q_2+\Gamma^T S(k+1)\Gamma]^{-1}\Gamma^T S(k+1)\Phi \qquad (6.15)$$

$$S(k)=\Phi^T S(k+1)\Phi+Q_1-L(k)^T[Q_2+\Gamma^T S(k+1)\Gamma]L(k) \qquad (6.16)$$

$$S(N)=Q_0 \qquad (6.17)$$

When N goes to infinity, we will have the steady state solution to the LQ problem. In the practical situation it is important that the controller should work for an undetermined future and not stop after N steps. Furthermore, the control algorithm should be simple to implement. Therefore, it is suitable to use the steady state solution of the LQ controller; $L(k) \rightarrow \bar{L}$ and $S(k) \rightarrow \bar{S}$ when $k \rightarrow \infty$. These are obtained from the solutions of the equations:

$$\bar{L} = \left[Q_2 + \Gamma\bar{S}\Gamma\right]^{-1}\Gamma^T\bar{S}\Phi \qquad (6.18)$$

$$\bar{S} = \Phi^T\bar{S}\Phi + Q_1 - \bar{L}^T\left[Q_2 + \Gamma^T\bar{S}\Gamma\right]\bar{L} \qquad (6.19)$$

Consequently equation (6.14) is modified to equation (6.20).

$$\tilde{u}(k) = -\bar{L}\tilde{x}(k) \qquad (6.20)$$

The control action on the process will be the sum of a feedforward steady state signal and the feedback signal as the solution of stationary LQ-regulator problem. The total control signal is computed as.

$$u(k) = u^{o} + \tilde{u}(k) \qquad (6.21)$$

At the beginning of this chapter, we discussed requirements for applications of the principle of gain scheduling to industrial plants. We stated that the pre-requisites for using the method are; the scheduling variable should vary slowly and the scheduling variable should capture the non-linearity of the process.

The first condition, that of slowly varying scheduling parameters is a serious restriction. Many processes are influenced by exogenous parameters which vary both slowly and rapidly. A gain scheduling strategy is presented for linear parameter varying systems by Shamma and Athans (1992). They give an approach to how to deal with fast changing scheduling variables by modifying the manner in which the operating point gains are scheduled. That is, the fixed point operating point gains should be scheduled in a manner which explicitly addresses the possibility of rapid variations of the process. Such an approach has the advantages that the closed loop stability is maintained and the fixed point properties remain. Although, the method is deducted to linear processes, it has been shown by an example that it is possible to use the same framework for non-linear systems.

The second condition, that the scheduling variable should capture the non-linearity of the process can be handled by specifying the stationary points of the process and computing a set of the auxiliary inputs which give the desired equilibrium states. Since computers have become cheaper and the computation speed has increased, it is possible to calculate the control signal for every combination of the exogenous signal at each sample. For this case, we do not use a look up table to find the gains; instead the control signal is computed on-line.

A practical way to handle the problem of rapidly varying scheduling parameters is to filter the stationary point signal with a low pass filter at each sample instance given by equations (6.22) and (6.23).

$$x_f^o(k+1) = \eta x_f^o(k) + (1-\eta)x^o(k) \tag{6.22}$$

$$u_f^o(k+1) = \eta u_f^o(k) + (1-\eta)u^o(k) \tag{6.23}$$

where $x_f^o(k)$ and $u_f^o(k)$ are the filtered stationary points. The filter constant is $\eta = e^{-h/T_f}$, with the sample time h and filter time constant T_f.

The solution of the LQ-regulator problem will be a time variable stationary solution on the form:

$$\bar{L}(k) = \left[Q_2 + \Gamma(k)\bar{S}(k)\Gamma(k)\right]^{-1}\Gamma(k)^T\bar{S}(k)\Phi(k) \tag{6.24}$$

$$\bar{S}(k) = \Phi(k)^T\bar{S}(k)\Phi(k) + Q_1 - \bar{L}(k)^T\left[Q_2 + \Gamma(k)^T\bar{S}(k)\Gamma(k)\right]\bar{L}(k) \tag{6.25}$$

where the gradients are computed at each sample instance:

$$\Phi(k) = \left.\frac{\partial F}{\partial x}\right|_{x_f^o(k),u_f^o(k)} \tag{6.26}$$

$$\Gamma(k) = \left.\frac{\partial F}{\partial u}\right|_{x_f^o(k),u_f^o(k)} \tag{6.27}$$

Consequently, the feedback signal is computed with a time variable matrix $\bar{L}(k)$ and equation (6.20) is modified to equation (6.28).

$$\tilde{u}(k) = -\bar{L}(k)\tilde{x}(k) \tag{6.28}$$

where we define case $\tilde{x}(k) = x(k) - x_f^o(k)$.

The total control signal manipulating the process is given by:

$$u(k) = u_f^o(k) - \bar{L}(k)\tilde{x}(k) \tag{6.29}$$

6.3 Application of the combined controller

6.3.1 Adaptation to the rinsing process

The principle of the construction of the control system is shown in Fig. 6.3. The process is controlled with the signal u(k) and the behaviour of the system is measured with the output signal y(k). These signals are inputs to the estimation block, which estimate both the states and the unknown parameters of the process. The stationary point is computed by the feedforward block. The stationary points are computed by the feedforward block and the feedforward signal is achieved by using a low pass filter. The feedback signal is computed are computed by the feedback block. Finally the feedforward and feedback signals are summed and fed into the process. It should also be pointed out that the four blocks are all influenced by the production parameters; strip velocity, strip width and strip thickness.

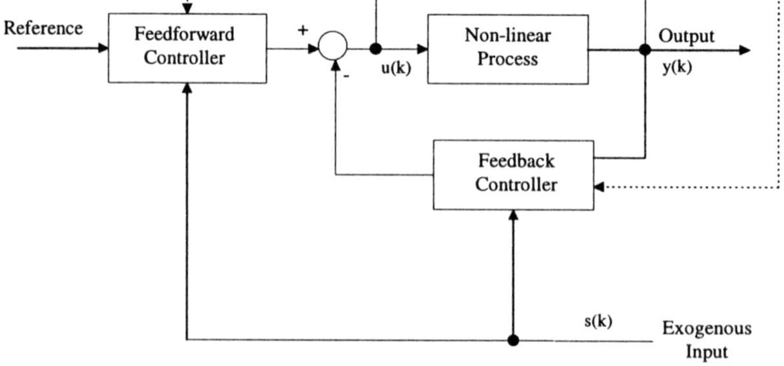

Fig. 6.3 Block diagram of the combined control system.

On-line estimation

The model used when the control signal is computed requires estimation of the unknown parameters and the states of the process. The estimation is based on the Extended Kalman Filter and is made in the same way as in Chapter 5. We use the same initial values and the same values of the covariance matrices.

Feedforward controller

Changes in the production variables influence the acidity of the rinse water. These production variables are measured when the steel strip passes the rinsing process. If the process is controlled only by a feedback controller, the controller compensates for the change in the production variables when they have influenced the concentration of the rinse water. Since the dynamics of the rinsing process are slow, it takes a long time for the controller to compensate for the changes in the process output. By means of the feedforward controller it is possible to compensate for the changes in the production variables at the moment the changes are detected and make the control system more efficient.

The steady state control signal is computed as the signal needed to achieve a desirable steady state concentration in tank 5. The control signal is based on the production variables, the condition of the squeezer rolls, and the reference value of the concentration in rinse tank 5.

The problem is formulated as finding the steady state solution of equation (6.30). Neglecting the disturbances acting on the states, we have to find the steady state vector x^o and control signal u^o which is given by equation (6.31). The steady state vectors depend also on the time variable s(k) and θ(k). Consequently, x^o and u^o will not be constants.

$$x(k+1)=F[u(k),s(k),\theta(k),x(k)]+v(k) \tag{6.30}$$

$$x^o(k)=F[u^o(k),s(k),\theta(k),x^o(k)] \tag{6.31}$$

where:

$$x^o(k)=[x_1^o, x_2^o, x_3^o, x_4^o, x_{5ref}]^T$$

$$u^o(k)=[F_{c1}]$$

$$s(k)=[B_v\ B_b\ B_t]^T$$

The parameters x_1^o to x_4^o are the steady state concentrations in rinse tank 1 to rinse tank 4. These values are used as references of the concentrations for the feedback controller. Furthermore, x_{5ref} is the reference concentration in tank 5, and u^o is the corresponding steady state control signal.

The control signal F_{c2} is not used by the controller. As before, F_{c2} is only for abnormal situations. A simulated control of the process with both F_{c1} and F_{c2} shows that the result is not better when both control signals are used. This can be explained by the fact that F_{c1} influences the main flow through rinse tank 5 to rinse tank 2, while F_{c2} only influences the main flow through rinse tank 2.

Since the process dynamics are slow and the production variables change frequently, it will take a long time before the steady state is achieved in the rinse tanks. Furthermore, strip velocity changes so often that the steady state values based on this production variable will almost never occur. To handle this problem, we filter the steady state values of the concentrations by a low pass filter, because they are used as reference values by the feedback controller. The time constants for the filter are chosen with roughly the same value as the time constants as for the rinse tanks. Rinse tank 1 has a time constant of about twenty hours because the flow of rinse water through this tank is low. The other tanks have a time constant of about two hours based on a main flow of 3.0 [m^3/h].

Since the strip velocity varies frequently, it follows that u^o will vary at the same rate. There is no point in changing the control signal so often, because the dynamic of the rinsing process is slow. Therefore, u^o is also filtered by a low pass filter. The strip thickness has a large influence on the concentration in the rinse tanks when the squeezer rolls have a surface of hard rubber. This means that the control signal must be able to follow the change in strip thickness. Consequently, we have to compromise between eliminating the high frequencies of the variations in strip velocity and following the changes in strip thickness. The time constant is set to 0.5 hours for the low pass filter of the control signal. This value of the time constant has also been verified from experiments with the process.

Feedback controller

From the discussion in Section 6.2, we also need a feedback controller because the feedforward controller cannot compensate for unpredictable disturbances. Furthermore, the feedforward controller cannot be perfect, and we also simplify by using a static feedforward controller. To take care of these defects we need a feedback controller.

Consider again the non-linear model of the rinsing process:

$$x(k+1)=F[x(k),u(k),s(k),\theta(k)] \quad (6.32)$$

$$y(k)=G[x(k)] \quad (6.33)$$

To develop the feedback controller, we apply a time variable stationary LQ-controller as described in Section 6.2. The linearized model is given by equation (6.34). As stationary values, we use the filtered steady state values formulated by equation (6.22):

$$\tilde{x}(k+1) = \Phi(k)\tilde{x}(k) + \Gamma(k)\tilde{u}(k) \quad (6.34)$$

where $\tilde{x}(k)$ is the deviation from the filtered stationary values of x(k) and $\tilde{u}(k)$ is the deviation from filtered the stationary value of the u(k). The matrices $\Phi(k)$ and $\Gamma(k)$ are the gradients of F with respect to x(k) and u(k). The gradients are based on; the stationary values, the estimated parameter vector $\theta(k)$ and the auxiliary inputs s(k).

Since the process states are estimated by the Extended Kalman Filter, we apply a LQG controller to the linearized model of the process, (Åström and Wittenmark, 1997). The feedback signal is computed from the estimate $\hat{\tilde{x}}(k)$ of $\tilde{x}(k)$ as:

$$\tilde{u}(k) = -\bar{L}(k)\hat{\tilde{x}}(k) \quad (6.35)$$

6.3.2 Simulation of the control system

The purpose of the simulation is to study the ability of the controller to compensate for changes in production variables and influence of the disturbances. The construction of the control system is shown in Fig. 6.3. As a "process", we use a "true model" of the process, which contains more estimated parameters than are used by the controller.

The difference between the two models is the determination of the flow via the strip. This flow is modelled by equation (4.15).

$$F_{bi}=K_{bi}\cdot B_v\cdot B_b+K_{ti}\cdot(B_t-B_{toffi})\cdot\alpha_i \tag{6.36}$$

$\alpha_i=1$ if $B_t > B_{toffi}$ otherwise $\alpha_i=0$

Note that equation (6.36) consists of three unknown parameters K_{bi}, K_{ti} and B_{toffi} for each pair of squeezer rolls. These parameters are identified off-line by optimising the likelihood function, see Chapter 4. During the simulation we use a "true model" that contains all three estimated parameters for each pair of squeezer rolls. However, the controller is based on a reduced number of unknown parameters, which are estimated on-line, see Chapter 5.

The steel strip is considered well rinsed, if the conductivity in rinse tank 5 is below 1.0 [mS/m]. This value corresponds to a scaled concentration in rinse tank 5 of 1.0. We assume that the scaled concentration should be below 1.0 with the probability of 0.975. Note, that the probability is based only on being below an upper limit. The reference value of the concentration in rinse tank 5 is then computed from equation (6.37), (Chung, 1979).

$$x_{5ref}=1-2\sigma \tag{6.37}$$

where σ is the standard deviation of the concentration in rinse tank 5.

To achieve a safe control system, the control signal has to be above a lower limit. The lowest flow of clean water into rinse tank 5 is limited to 1.0 [m^3/h].

The simulation is done with one of the sequences of the production variables described in Section 5.2. The corresponding estimates of K_{bi}, K_{ti} and B_{toffi} obtained from off-line parameter identification are used to simulate the "process". Process noise and measurement noise are discrete time Gaussian white noise with zero mean and covariance matrices R_1 and R_2. The values of R_1 and R_2 are the same as used in Chapter 4. The simulation presented is done with the production variables from measurement sequence 1, see Chapter 5. Measurement sequences 2 and 3 give similar results.

The matrices Q_1 and Q_2 are chosen by repeated simulations. The deviation in rinse tank 5 should be punished harder than the deviation in rinse tank 1. The reasons are that the concentration in rinse tank 5 is most important and the concentration in rinse tank 1 has a much longer time constant than the other rinse tanks. A suitable choice of weights is:

Q_1=diag[0.1 1 1 1 10]

Q_2=1

- **Comments on the conductivity.**

  The same diagram shows both the output and the corresponding reference signal, (see Fig. 6.4). The conductivity in rinse tank 1 varies more slowly than the conductivity in the other rinse tanks. This agrees with the fact that the time constant for rinse tank 1 is longer than the time constants for the other rinse tanks. The conductivity in rinse tank 2 varies more. This is because the rubber of the squeezer rolls before rinse tank 2 is harder than the other squeezer rolls, and the flow beside the strip is large. This means that the flow from rinse tank 1 increases when the thickness of the strip increases.

  The conductivity in tank 5 is mostly below the upper limit of 1.0. The reference value of the concentration is computed from equation (6.37), and we use a value of the standard deviation of σ=0.10. This is achieved from a previous simulation with a reference scaled concentration of 1.0. With this value of σ we achieve a reference concentration of 0.80. From the simulation we have that the mean value of the conductivity in rinse tank 5 is 0.79.

- **Comments on the estimated parameters.**

  The values of the estimated parameter vector $\theta(k)$ are about the same as achieved in Chapter 5, see Fig. 6.5 and Fig. 6.6.

- **Comments on the control signal.**

  We can note from the diagram, that the lower limit of the control signal is not a serious restriction, see Fig. 6.7, since the control signal seldom reaches the value of 1.0 [m^3/h]. There is also a margin for the control signal to reach the maximum value of 6.0 [m^3/h]. The control signal varies moderately, which means that the variations of the strip velocity do not fully influence the control signal.

If we let the time constant of the filter for the control signal be too long, there will be a problem to compensate for dependence of strip thickness of the squeezer rolls at the end of the rinsing process. This means that the feedforward controller is not fast enough to compensate for the variation of the strip thickness. These variations will instead be compensated by the feedback controller, which needs longer time because this part of the controller is dependent on the occurrence of a deviation in the concentration in the rinse tanks. If, instead, we let the time constant for the filter be shorter, the feedforward controller will react faster to the variations of the production variables. As result of repeated simulations, we have chosen a suitable compromise of the time constant of the filter for the control signal of 0.5 hours. With this filter constant, the feedforward controller will compensate for variation in the production variables, and the control signal will not vary too much.

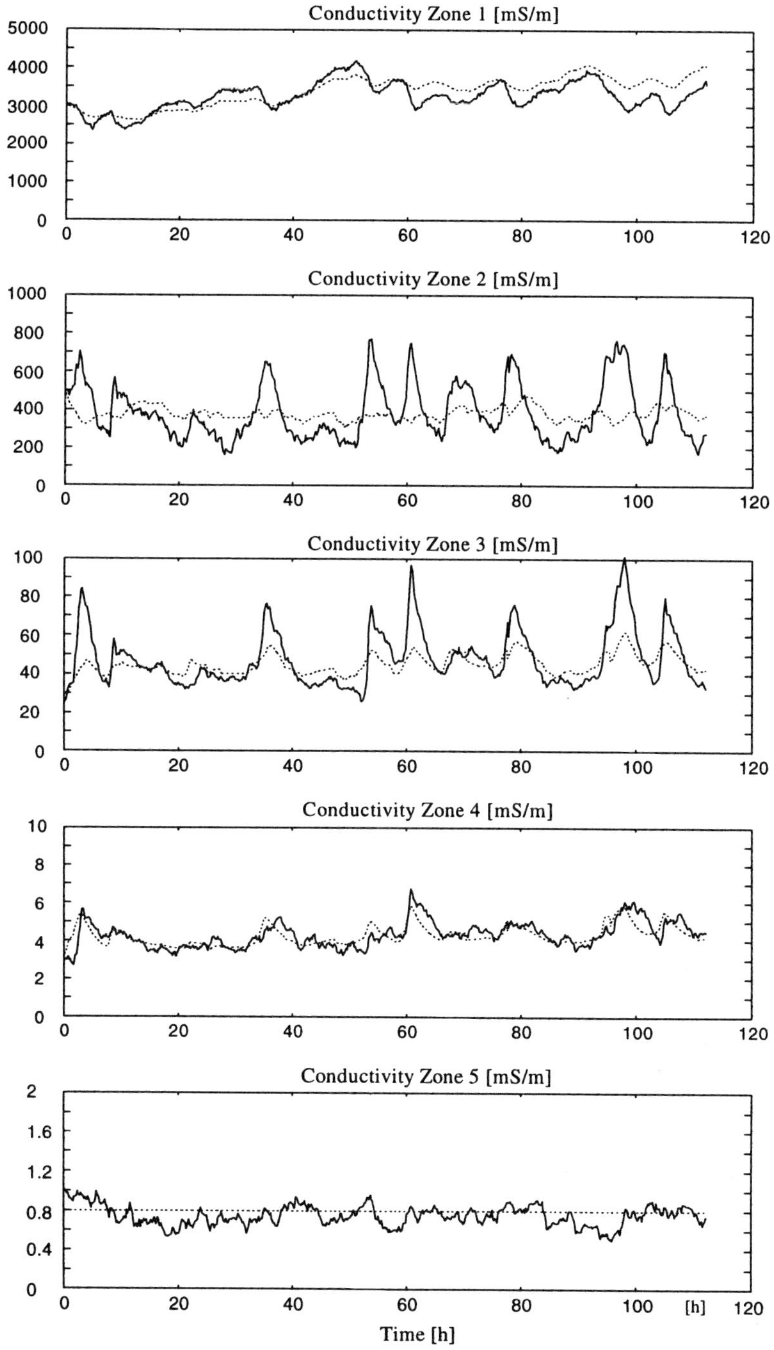

Fig. 6.4 Simulated outputs, *dotted* - reference.

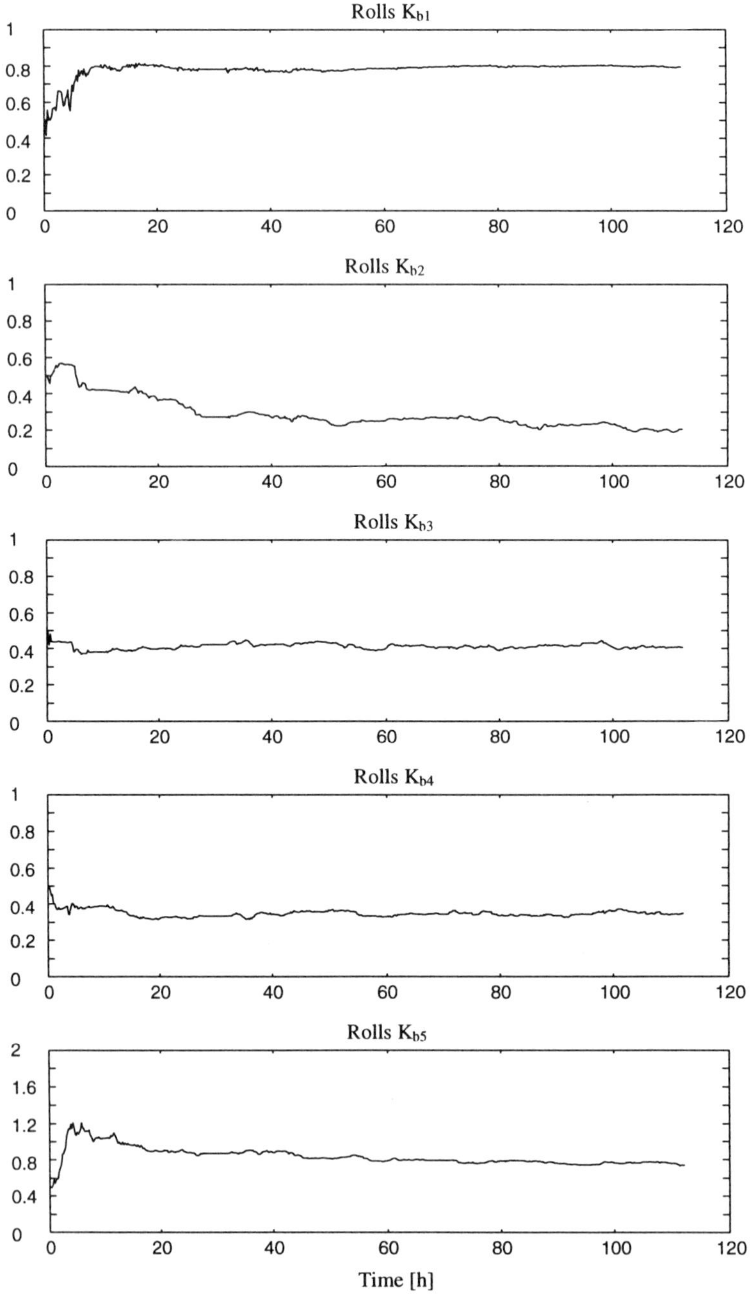

Fig. 6.5 Estimated parameters from simulation (K_{bi}).

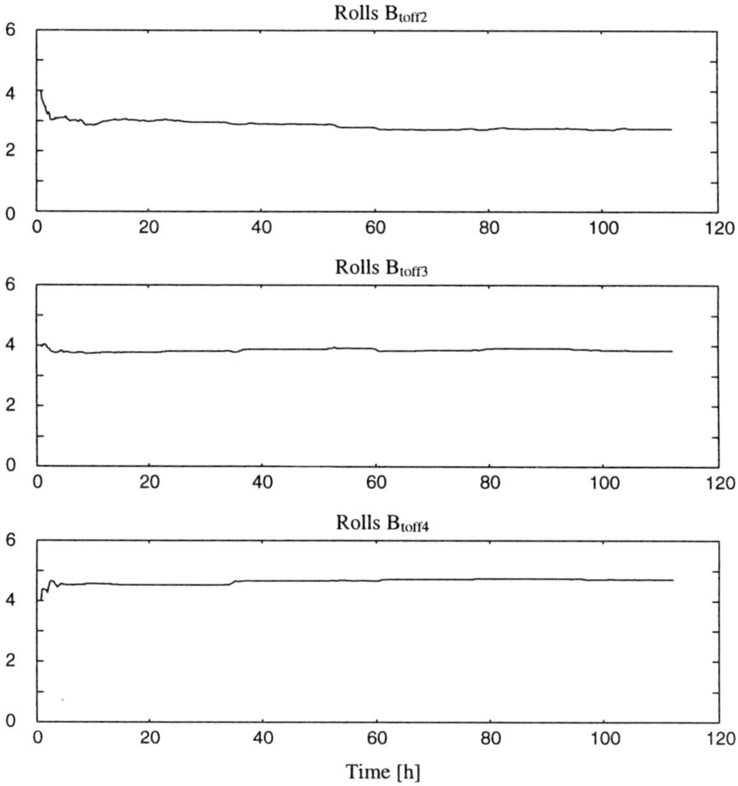

Fig. 6.6 Estimated parameters from simulation, (B_{toffi}).

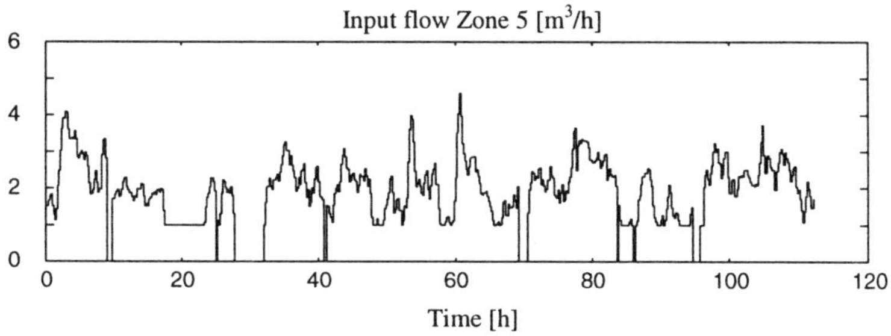

Fig. 6.7 Control input signal from simulation, ($u=F_{c1}$).

6.3.3 Control of the process

The construction of the control system is the same as in Fig. 6.3. The block "process" is now the rinsing process. Estimation of unknown parameters and computing of the control signal is carried out the in same as during simulation of the control system.

The rinsing process is tested with the combined controller during several occasions. This means that the controller is examined on several different conditions of the squeezer rolls. A result of these tests is that the filter constant for filtering the feedforward control signal must not be too long. This fact was also verified during simulation of the control system. If a pair of the squeezer rolls at the end of the rinsing process is hard, the flow beside the strip is much dependent on the strip thickness. The result of the controlling the process is then much better if the filtering constant is decreased from 2.0 hours to 0.5 hours, because the controller will react much faster to changing strip thickness.

To compute the reference concentration in rinse tank 5, we need an estimation of the standard deviation of the concentration in rinse tank 5. From the different tests with the combined controller a reasonable estimation of the standard deviation is about σ=0.1.

Fig. 6.8 shows a result from the control of the process lasting seven days. The results shown are typical of this method of controlling the process.

- **Comments on the conductivity.**

  The conductivity in rinse tank 1 is higher than the results shown in Chapter 4. This is typical of the production situations when the control experiments were carried out. The conductivity in rinse tank 5 is mostly below the upper limit. The mean value of the conductivity in rinse tank 5 is 0.77 and the standard deviation is σ=0.12. This result could be compared with the fact that the controller is based on a standard deviation of 0.1 and consequently a reference value of 0.8.

- **Comments on estimated parameters.**

  From Fig. 6.9 it is seen that the squeezer rolls 2 and 3 have about the same efficiency. The flow via the top and bottom of the strip is small. The thickness of the strip has very little influence on the flow beside the strip, see Fig. 6.10. Note, that the squeezer rolls 2 are the last pair of squeezer rolls to be replaced. They were replaced the day before the experiment started.

The pair of squeezer rolls 4 is, however, in bad condition; the flow via the top and bottom of the strip is more than double that of the flow after squeezer rolls 2 and 3, see Fig. 6.9. We can see from Fig. 6.10, that B_{toff4} is about three. This means, that the rubber of squeezer rolls 4 is hard and the flow beside the strip is large.

For squeezer rolls 1 and 5, only one parameter each is estimated. Squeezer rolls 5, have about the same efficiency as squeezer rolls 2 and 3.

❐ **Comments on the control signal.**

When the steel strip is stopped, the input flow of clean water is turned off. From the figures it is shown that there is more clean water available for use if necessary. It can also be noted that the variation in the control signal is moderate. The mean consumption of clean water while the process was running is 1.3 [m^3/h].

❐ **Comments on the production variables.**

The production variables are typical of the mix of steel dimensions and the way the process runs, see Fig. 6.12.

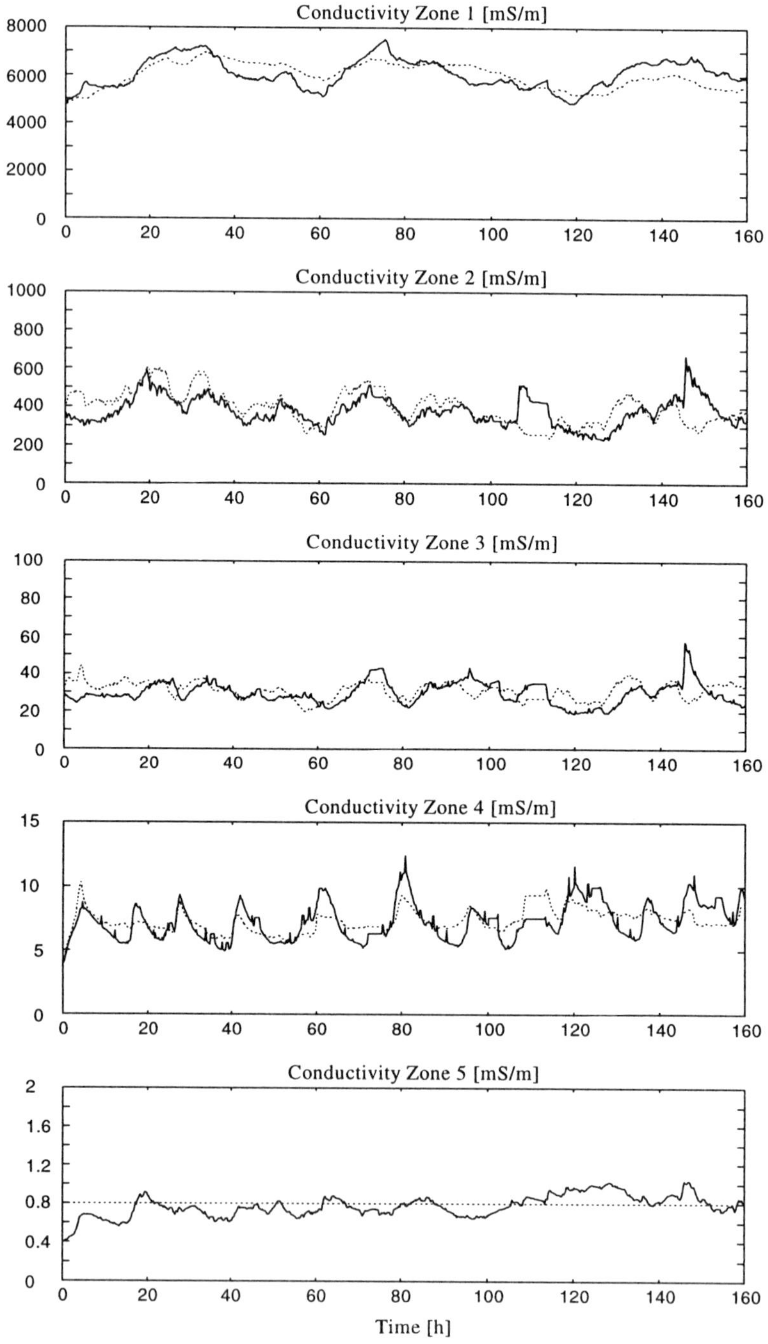

Fig. 6.8 Process outputs, *dotted* -reference.

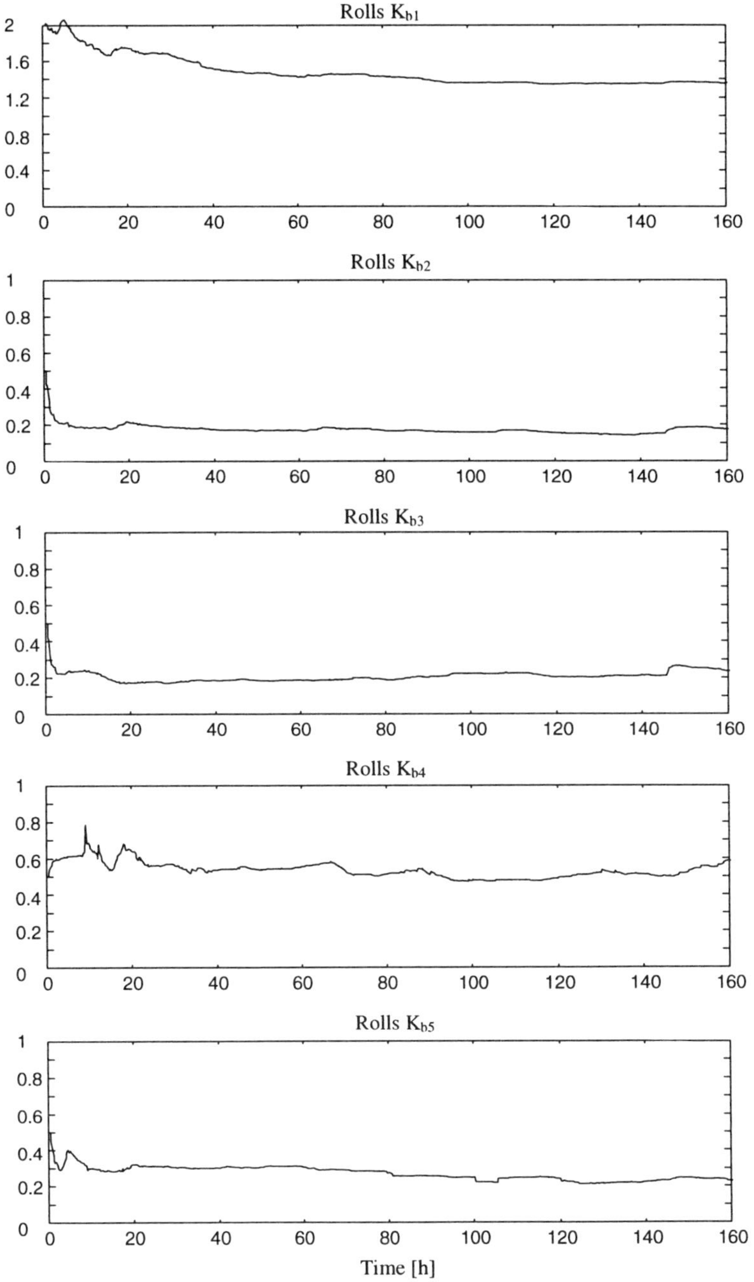

Fig. 6.9 Estimated parameters from control of the process, (K_{bi}).

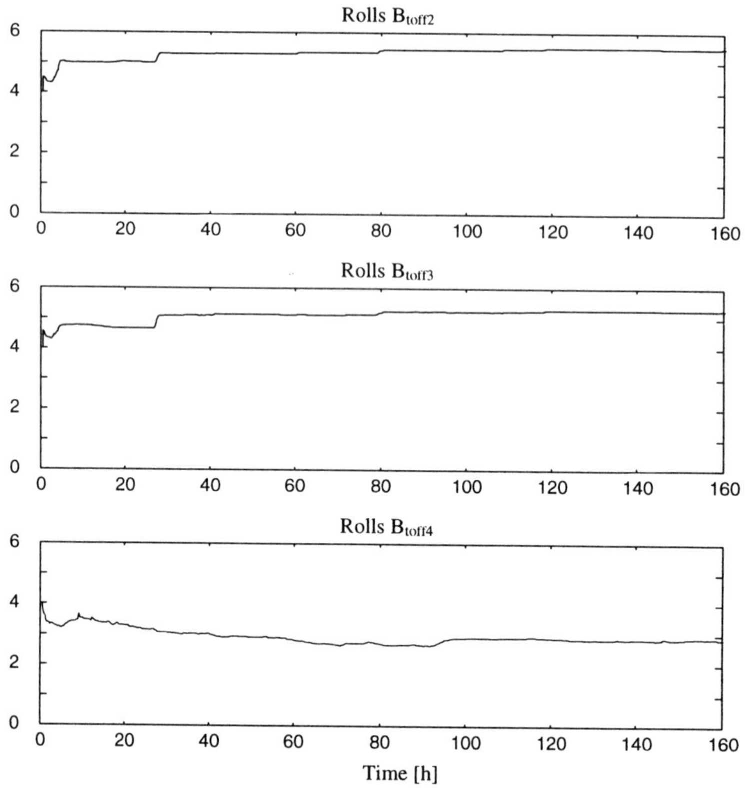

Fig. 6.10 Estimated parameters from control of the process, (B_{toffi}).

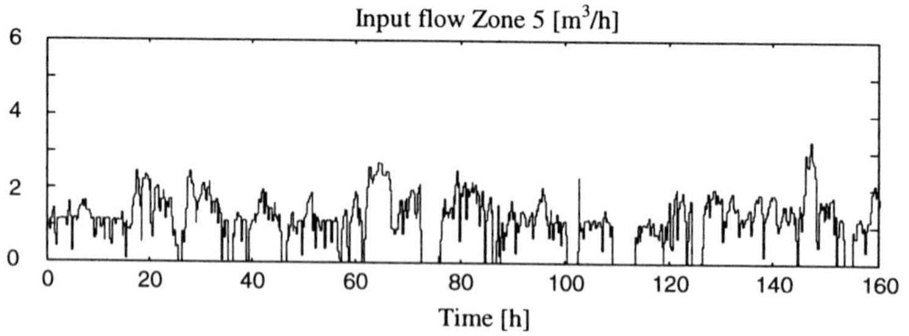

Fig. 6.11 Control signal from control of the process, ($u=F_{c1}$).

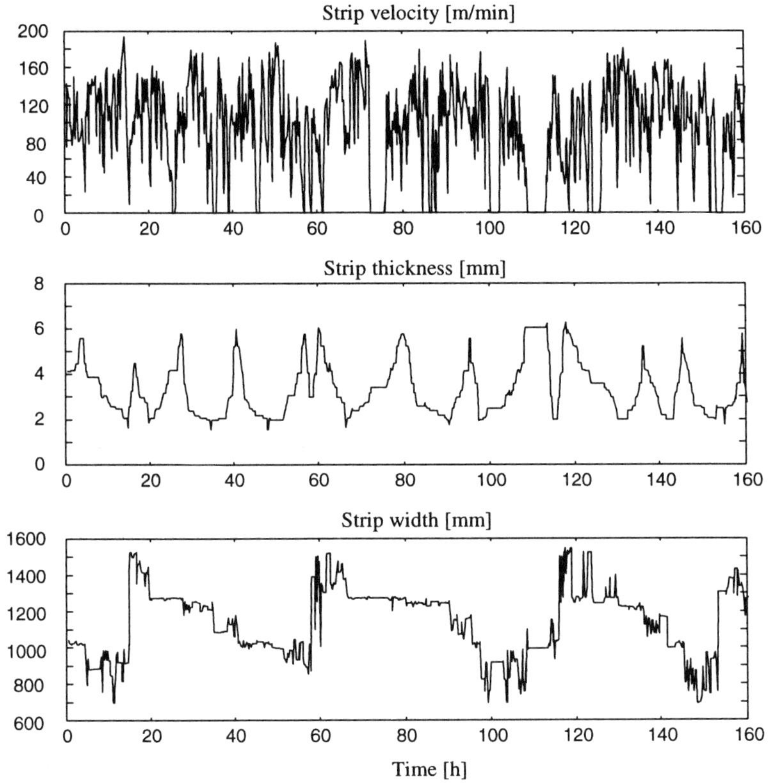

Fig. 6.12 Auxiliary inputs from control of the process, the production variables.

6.4 Non-linear optimal controller

The computation capacity of work stations has been considerably increased in the last few years. In that way it has become interesting to apply advanced control theory to industrial processes. However, these methods have a much longer history. Kalman presented in the early 1960's a work where he sought to determine when a linear control system can be said to be optimal, (Kalman, 1960). Kalman studied so-called Linear Quadratic Regulator designed to minimise a quadratic performance index.

An elegant mathematical solution to the model based optimal control problem was presented by Pontryagin, which was called the Pontryagin minimum principle, (Pontryagin et. alt 1962). An improvement to make the control problem more realistic was to introduce physical limits on the control signals and process states. The approach of Pontryagin leads to a non-linear two point boundary problem, which, computationally is a demanding task.

The method of dynamic programming was developed by Bellman in the mid 50's, (Bellman, 1957) and later on applied to control theory, (Bellman and Kalaba, 1965). Bellman introduced the "principle of optimality" and extended it to a sequence of decisions which formed an optimal solution.

Since the huge increase in computation capacity, the number of optimal control applications to industrial processes is considerable. However, many applications have revealed the gap between theory and practice, and the control applications are often strongly process dependent.

It is desirable to form a framework to carry out an applicable optimal control strategy. It is a fact that only the future behaviour of a process can be controlled. The present state of a system is the end of the past. The future behaviour is dependent on the present state, the future control signals, and other measurable and unmeasurable variables.

When a process is non-linear and both control and state variables are constrained by physical reasons, a state space model is useful to describe the process. Another reason for state space modelling is when the system is affected with other variables than for control purposes. Furthermore, when a state space model is applied in a general optimisation scheme, it is possible to determine the future control actions which are optimal in terms of any performance criterion, (Balchen et. alt, 1992).

In the steel industry there are many processes which can be modelled by state space models and can be manipulated by an optimal predictive controller. For example; for reheating furnaces, the model is based on physical laws and a heuristic search algorithm is used to determine the zone temperature, (Yang and Lu, 1988). Models describing the strip thickness during hot strip rolling and the control possibilities is presented by Kato et. alt (1992). A state space model of the tension in a bar rolling mill is developed by McFarlane and Stone (1990). The list can be made long, but these are only some examples which describe the potential in applying model based optimal predictive control system. For further examples of applications within the steel industry, (Sohlberg, 1993).

To formulate the optimal control problem for such a system the following are needed, (Kirk, 1970):

- A mathematical model of the process to be controlled.
- Concretization of the physical limits of the process.
- Specification of a performance measure.

Mathematical model

One of the difficult parts of the optimal control problem of non-linear processes is to develop an adequate process model, which is able to predict the behaviour of the system based on all relevant process inputs. In Chapter two, we have presented a method to generate non-linear state space models based on a priori knowledge of the system. To simplify the presentation, we restrict the model to be described with discrete time equations:

$$x(k+1) = F\left[x(k), u(k), s(k)\right] + v(k) \tag{6.38}$$

$$y(k) = G\left[x(k), u(k), s(k)\right] + w(k) \tag{6.39}$$

Physical limits

The constraints of the process consist of the range of the actuators and sensors. Furthermore, the process states or process outputs are constrained or limited, depending on the future behaviour of the process. Even the change in the control signal can be constrained due to the desired behaviour. The constraints for the control signal and states can be formulated:

$$u_{min} \leq u(k+i) \leq u_{max} \qquad (6.40)$$

$$x_{min} \leq x(k+i+1|k) \leq x_{max} \qquad (6.41)$$

where u_{min} and u_{max} are the constraints for the control signal, x_{min} and x_{max} are the constraints for the process states. The variable p denotes the future increment index, i=0, 1, ..., m, and m is the prediction horizon.

Performance measure

In order to find an optimal controller, a performance measure is needed. This reflects the goal of controlling a specific industrial process. It is well known that optimal control is not a specific problem; instead the control action is optimal in some sense.

A common general form of the permanence measure is formulated as:

$$J = \varphi[x(m+1)] + \sum_{k=i}^{m} L[x(k), u(k), k] \qquad (6.42)$$

where φ and L are scalar functions.

The performance measure can be formulated more specifically suitable for optimal control of state space models:

$$J = \|x(m+1) - x_r(m+1)\|_{Q_m}^2 + \sum_{k=i}^{m} \left[\|x(k) - x_r(k)\|_{Q_1}^2 + \|u(k)\|_{Q_2}^2 \right] \qquad (6.43)$$

where $\|x(k) - x_r(k)\|$ is the norm of the vector $[x(k) - x_r(k)]$.

The indices Q_m, Q_1 and Q_2 are real symmetrical positive semi-definite matrices, which are used as weighting matrices. The first term in equation (6.43) penalises the deviation of the final state from its desired value $x_r(m+1)$. The second term penalises the deviation of the state and the third term the control effort during the predicted interval.

Based on the mathematical model and the physical constraints, we search for a set of control signals u(k), u(k+1), ..., u(k+m), which minimises the performance measure given by equation (6.43). The optimal control problem is illustrated by Fig. 6.13.

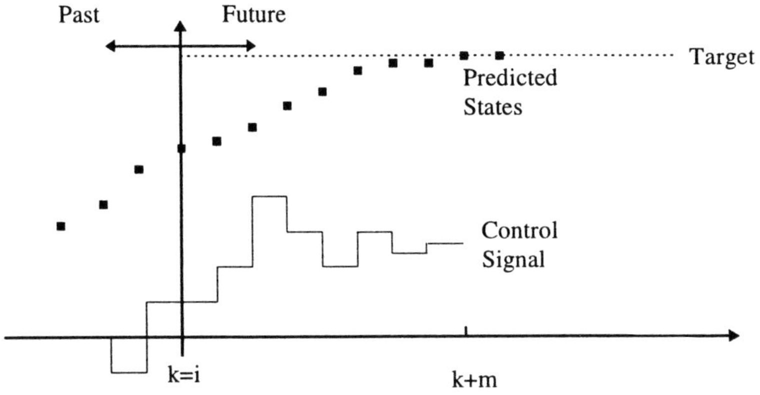

Fig. 6.13 Model based optimal control

The minimisation of the performance measure J, is here based on the Sequential Quadratic Programming (SQP-method), (Powell, 1978). A routine for solving the optimisation problem with SQP is available in Matlab Optimisation Toolbox, (Grace, 1990), and is used with Matlab™.

Generally, the routine finds the constrained minimum of a scalar function f(X), of a vector X, starting at an initial estimate. The minimisation is subject to given constraint, G(X)=0. This problem is generally referred to as constrained non-linear optimisation, and is mathematically expressed as:

$$\underset{X}{\text{Minimise}}\ f(X) \quad \text{subject to a constraint } G(X) = 0 \tag{6.44}$$

Applying the SQP-method to the formulated optimal control problem, we achieve an attractive formalism. The final state is not involved in the permanence measure since the optimisation is repeated for every new sample and only the first control action u(k) is implemented. The reason is that typical industrial processes are influenced by changing auxiliary inputs and time-varying model parameters. The constrained optimal control problem is formulated by equation (6.45).

$$\underset{[x(i),\dots,x(m),u(i),\dots u(m)]}{\text{Minimise}} \quad J = \sum_{k=i}^{m}\left[\|x(k) - xr(k)\|_{Q_1}^{2} + \|u(k)\|_{Q_2}^{2}\right] \qquad (6.45)$$

subject to dynamic constraint: $x(k+1) - F\left[x(k), u(k), s(k)\right] = 0$

with the limits: $u_{min} \leq u(k) \leq u_{max}$

$$x_{min} \leq x(k) \leq x_{max}$$

❐ **Remark**

The non-linear optimal control system is closely related to a general Model Predictive Controller (MPC), (Camacho and Bordons, 1995). The MPC system is not a specific control strategy but more a range of control methods developed around certain ideas. The main differences between the non-linear optimal controller presented here and the general MPC is that, for the MPC, two predictive horizons can be used, one for the control signal and other for the process output. Furthermore, the performance measure consists of the deviations between the process output and the reference of the output, and the control increments.

6.5 Application of the optimal controller

6.5.1 Adaptation to the rinsing process

The optimal control system has two different parts in addition to the process, see Fig. 6.14. One of the parts consists of estimation of the states and of the unknown parameters. This part is identical with the corresponding part of the combined control system, see Fig. 6.3. The other part of the control system consists of the optimal controller. In addition to the input and output for the blocks, they are influenced by auxiliary inputs, the strip velocity, strip width and strip thickness.

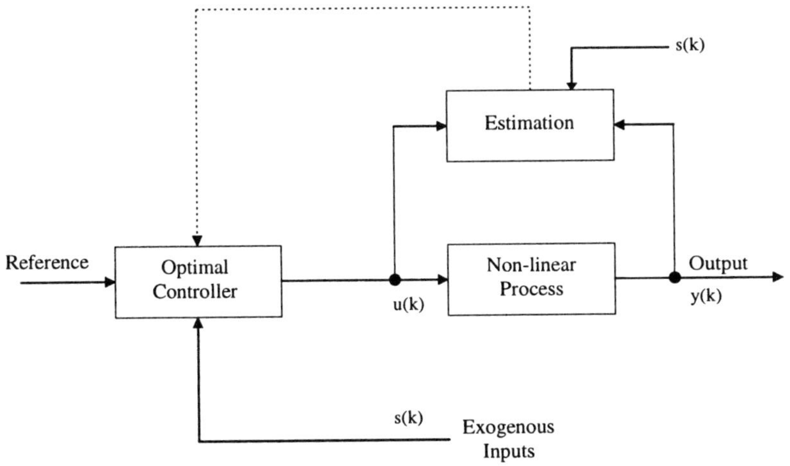

Fig. 6.14 Block diagram of the optimal control system.

On-line estimation

The model used when the control signal is computed, requires estimation of the unknown parameters and the states of the process. The estimation is based on the Extended Kalman Filter and is made in the same way as in Chapter 5. We use the same initial values and the same values of the covariance matrices.

Optimal controller

As mentioned earlier, we require a mathematical model of the process, the physical limits of the process variables and a loss function in order to be able to formulate the optimal control problem.

From Chapter 4, we have a mathematical model of the process. It is given in a discrete state space form, which is suitable to be used for non-linear optimal control of the process. The model we use also gives an adaptive description of the process, because the unknown parameters are estimated on-line. The model is given as:

$$x(k+1)=F\,[u(k),s(k),\theta(k),x(k)]+v(k) \qquad (6.46)$$

$$y(k)=G\,[x(k)]+w(k) \qquad (6.47)$$

The physical limits of the process consist mainly of the limits of the control signal. The control signal has an upper limit of 6.0 [m^3/h]; this is discussed further in Chapter 7. It also has a lower limit; for safety reasons we do not want to risk the liquid levels in any rinse tanks being lower than a specified value. The concentrations in the tanks are obviously positive. The specific process constraints are formulated as:

$$1.0 \le u(k) \le 6.0 \qquad (6.48)$$

$$0 \le x(k) \qquad (6.49)$$

The purpose of controlling the process is the same as in the case of the combined controller. This means that the scaled concentration in rinse tank 5 must be below 1.0 with a specified probability. In Section 6.2, the reference value is computed from equation (6.50).

$$x_{5r}=1-2\sigma \qquad (6.50)$$

where σ is the standard deviation of the concentration in rinse tank 5.

The control problem is solved by finding a control sequence which minimises the loss function J, given by equation (6.51). During minimisation the values of the production variables and estimated parameters are used as obtained at the sample instances k=i, because we do not have information about the future values.

The loss function is based on the concentration in rinse tank 5 and the flow of clean water into tank 5 and is formulated as:

$$J = \sum_{k=i}^{m} \left\{ Q_1 \left[x_5(k) - x_{5r}(k) \right]^2 + Q_2 \left[u(k) - u_r(k) \right]^2 \right\} \qquad (6.51)$$

The loss function is interpreted as a punishment of the deviation in the concentration in tank 5 from a reference value. We punish the deviation because it is a regulator problem to keep the concentration close to the reference value. The concentrations in the other rinse tanks are not considered in the loss function, because it is the acidity of the rinse water in rinse tank 5 which decides how clean the steel strip is when it leaves the rinsing process.

Note that if we only punish the magnitude of the control signal, we will have a bias during control of the process. This is a result of the fact that the loss function is a quadratic function of the control signal, thus large amplitude of the control signal is punished more than small amplitudes. Therefore, the deviation of the control signal is punished in the loss function. The reference value of the control signal is chosen as 2.0 [m^3/h], which is approximately the mean value of clean water feed into the process.

The minimisation of the loss function J is repeated at every new sampling instant. This means that the new values of the production variables and the estimated parameters are used at every new sample instant. This means also that the disturbances acting on the system are considered at every new sampling instant via the Extended Kalman Filter.

An alternative performance measure to minimise is given by equation (6.52). This function is a sum of the absolute values of the deviations from the reference values.

$$J = \sum_{k=i}^{m} \left\{ Q_1 \left| x_5(k) - x_{5r}(k) \right| + Q_2 \left| u(k) - u_r(k) \right| \right\} \qquad (6.52)$$

However, the minimisation routine is based upon functions with continuous first and second derivatives. The function given by equation (6.52) has discontinuous derivatives in the solution points. Consequently, this loss function will not be used.

We have made two separate routines in Matlab™, which are called by the optimisation routine. One routine computes the loss function and admissible states. The second routine computes the derivatives of the loss function and the function G(X) with respect to the control signal and the concentration in tank 5. The derivatives can be computed by the optimisation routine, but the speed of computation was increased when we made a separate routine.

6.5.2 Simulation of the control system

The purpose of the simulation is to study the ability of the optimal controller to compensate for variations in the production variables strip velocity, strip width and strip thickness. Furthermore, it is of interest to investigate whether the time for minimising the loss function is considerably shorter than the sample interval. As a "true model" we use the same model as during simulation of the combined controller. The same sequence of production variables is also used during the simulation with the optimal controller. Process noise and measurement noise are discrete time Gaussian white noise with zero mean and covariance matrices R_1 and R_2. The values of R_1 and R_2 are the same as used during the test of the Extended Kalman Filter.

The value of σ used in equation (6.50) is computed by first making the simulation with a reference value x_{5r}=1.0 and then computing the value of σ. Afterwards, the simulation is repeated with a new value of x_{5r} computed from equation (6.50) with the value of σ. The simulation is done with Q_2=1 and is tested with three different values of Q_1; 10, 100, 1000. The best result was achieved with Q_1=1000. However, the control signal will vary very much so we chose Q_1=100. The result of the simulation is shown in Fig. 6.15.

❒ **Comments on the conductivity.**

The value of the conductivity is mostly below the upper limit 1.0, see Fig. 6.15. The reference value of the concentration in rinse tank 5 is computed from equation (6.50) with σ=0.08. This value of the standard deviation gives a value of the reference concentration of 0.84. The simulation gives a mean value of conductivity in the rinse tank of 0.82. Note, that the optimal controller gives a somewhat better result than the combined controller.

❒ **Comments on the estimated parameters.**

The values of the estimated parameters are about the same as achieved during the simulation of the combined control system, see Fig. 6.16 and Fig. 6.17.

❒ **Comments on the control signal.**

From the figure it is shown that there is ample margin to the upper limit of the control signal, see Fig. 6.18. The control signal reaches the lowest limit during some shorter periods. During these periods the conductivity is decreased. However, the decrease is not large and the only influence of the process is that the strip will be cleaner than necessary The increase in consumption of clean water due to this limit is moderate.

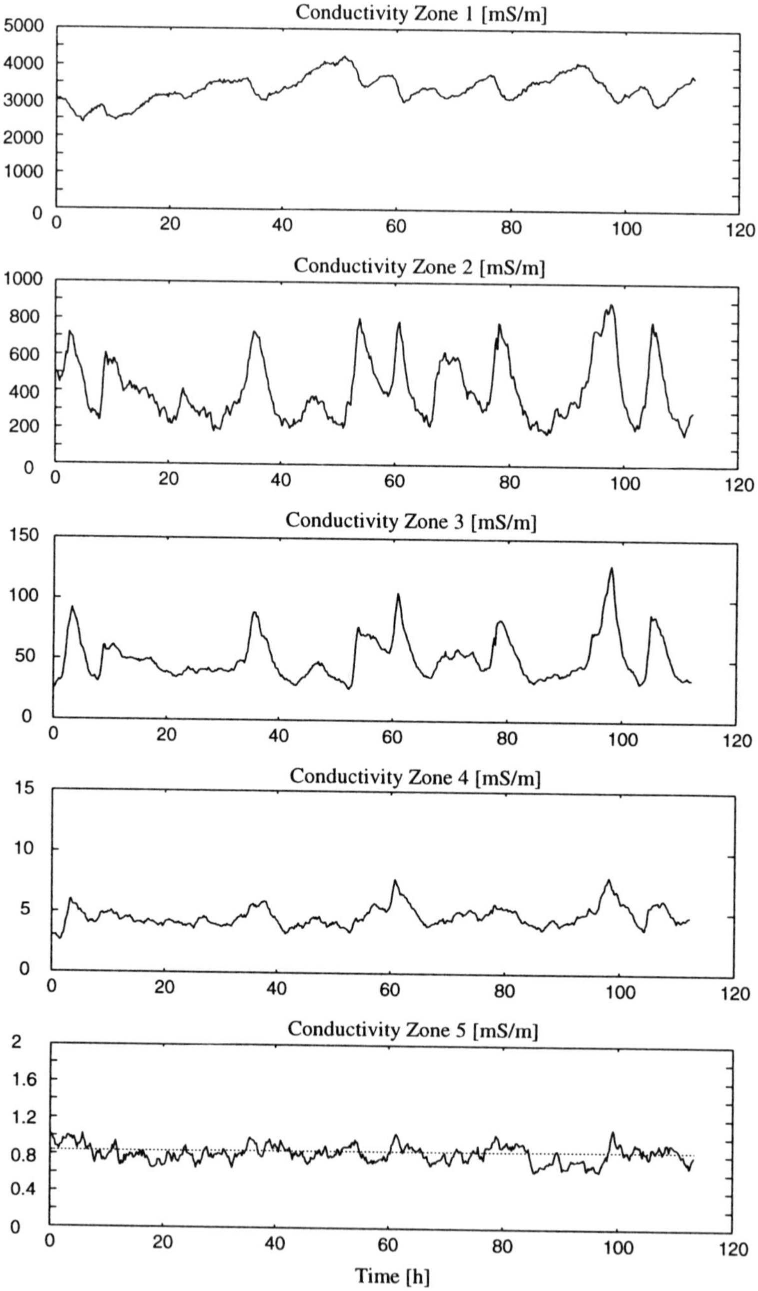

Fig. 6.15 Model outputs from simulation, *dotted* - reference.

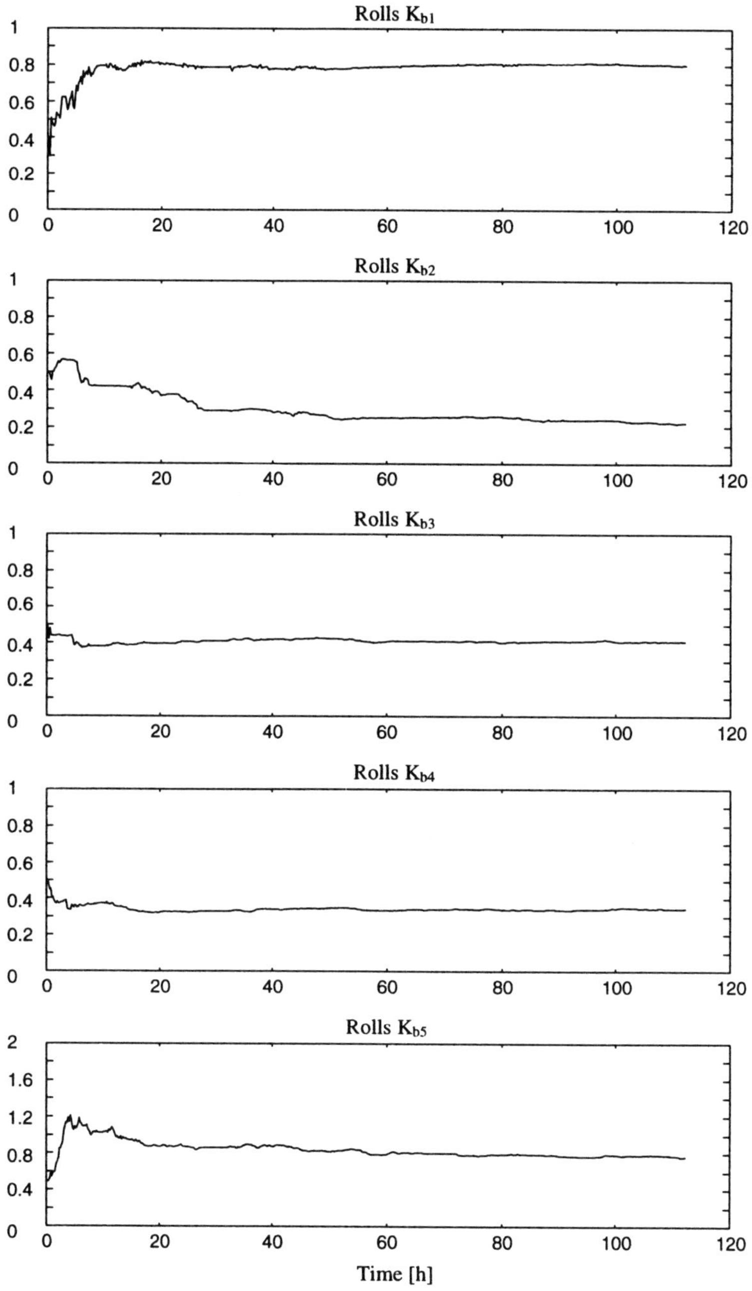

Fig. 6.16 Estimated parameters from simulation, (K_{bi}).

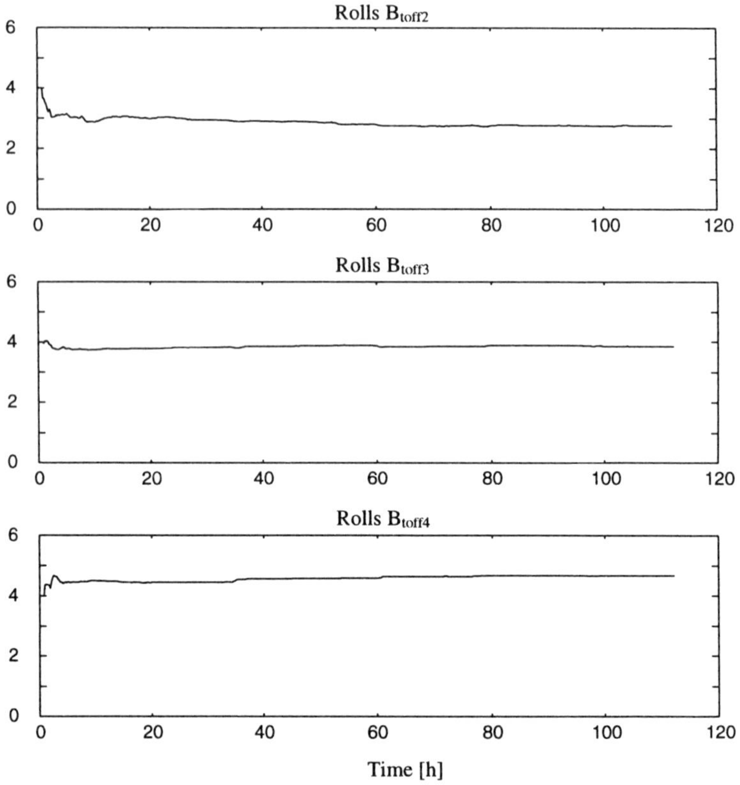

Fig. 6.17 Estimated outputs from simulation, (B_{toffi}).

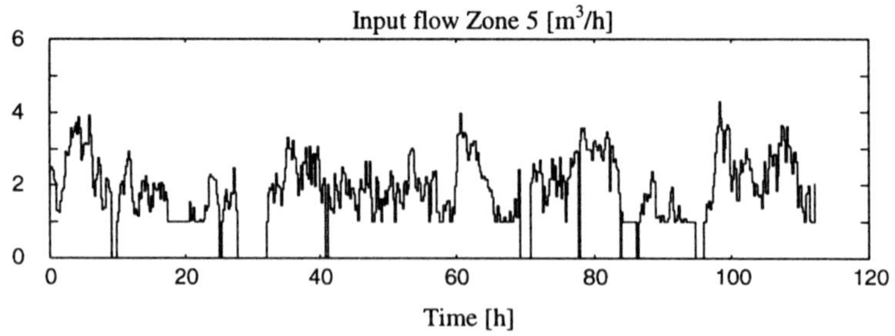

Fig. 6.18 Input signal from simulation, ($u=F_{c1}$).

6.5.3 Control of the process

The construction of the control system is the same as in Fig. 6.14. Estimation of the states of the process and unknown parameters is made in the same way as during the simulation, and we use the same limits of the control signal.

The process has been controlled by the optimal controller several times with varying values for Q_1. The tests give that suitable values of the weight constants are $Q_1=100$ and $Q_2=1$. A test with $Q_1=1000$ shows that the control signal will vary very much and be sensitive to disturbances and changes in production variables. This result agrees also with the simulated tests.

The reference concentration is computed from equation (6.50). As an initial value of the standard deviation we set $\sigma=0.1$ during the first two days of controlling the process. After two days σ is updated at every sample instance and a computation is made based on data from the last two days. The computation for σ had been based on a long period; otherwise σ would vary more and this would mean that x_{5r} would vary more.

The control of the rinsing process is shown in Fig. 6.19. The results are typical of the optimal controller based on $Q_1=100$. The process has been controlled during a period of seven days.

❑ **Comments on the conductivity.**

The conductivity value in rinse tank 1 varies moderately and slowly. The conductivity value in rinse tank 2 varies with higher frequencies. The conductivity value in rinse tank 5 is mostly below the specified limit of 1.0 [mS/m]. Note that the reference concentration in tank 5 is increased after 48 hours, because we start to update the standard deviation after two days. The reference value changes slowly to a higher value because the concentration in rinse tank 5 is low at the beginning of the control sequence. For the last 100 hours of the measurement sequence the mean value of the conductivity is 0.926 and the standard deviation is 0.045. The reference value computed from equation (6.50) is about 0.91.

From the estimation of the unknown parameters, we can see that the squeezer rolls used during the experiment are more efficient than the rolls from the simulation. This explains why the result from the experiment is better than the corresponding result from the simulation.

❐ **Comments on the estimated parameters.**

The efficiency of squeezer rolls 2, 3 and 4 is about the same. However, the pair of squeezer rolls 5 is not in as good condition as the other pairs of squeezer rolls, see Fig. 6.20 and Fig. 6.21.

❐ **Comments on the control signal.**

When the steel strip is stopped, the flow of clean water into tank 5 is turned off, see Fig. 6.22. The flow of clean water into rinse tank 5 varies moderately, which means that the choice of Q_l is made correctly. The figure shows that it is possible to increase the magnitude of the control signal. In Chapter 7, we discuss the change of the squeezer rolls in relation to the consumption of clean water. The mean flow of clean water into tank 5 was 1.5 [m^3/h] during the experiment.

❐ **Comments on the production variables.**

These variables are typical for the way production is planned and the process is run, see Fig. 6.23.

Applying control theory to an industrial process involves both theoretical and practical work. For safety reasons, it is necessary to be careful before a new controller is used. If the controller fails, there will be losses in production volume or losses in quality. Therefore, a new controller was first run parallel with the ordinary controller, without controlling the process. During these test periods, the unknown parameters were also estimated. Since the process operators were interested in the results of the tests, they had access to the information about which pair of squeezer rolls was in poorest condition. This information was used by the process operators to decide which pair of rolls should be replaced at the next planned stop. A result of this guidance is that the difference in the parameter estimate between the rolls has decreased. It can be seen in Fig. 6.20 and Fig. 6.21. that the difference between the estimated values of K_{b2}, K_{b3}, K_{b4} is small and the difference between B_{toff2}, B_{toff3}, B_{toff4} is also small. The interest in the efficiency of the rolls justified the estimation of the parameters also during the periods when there were no tests of a new controller.

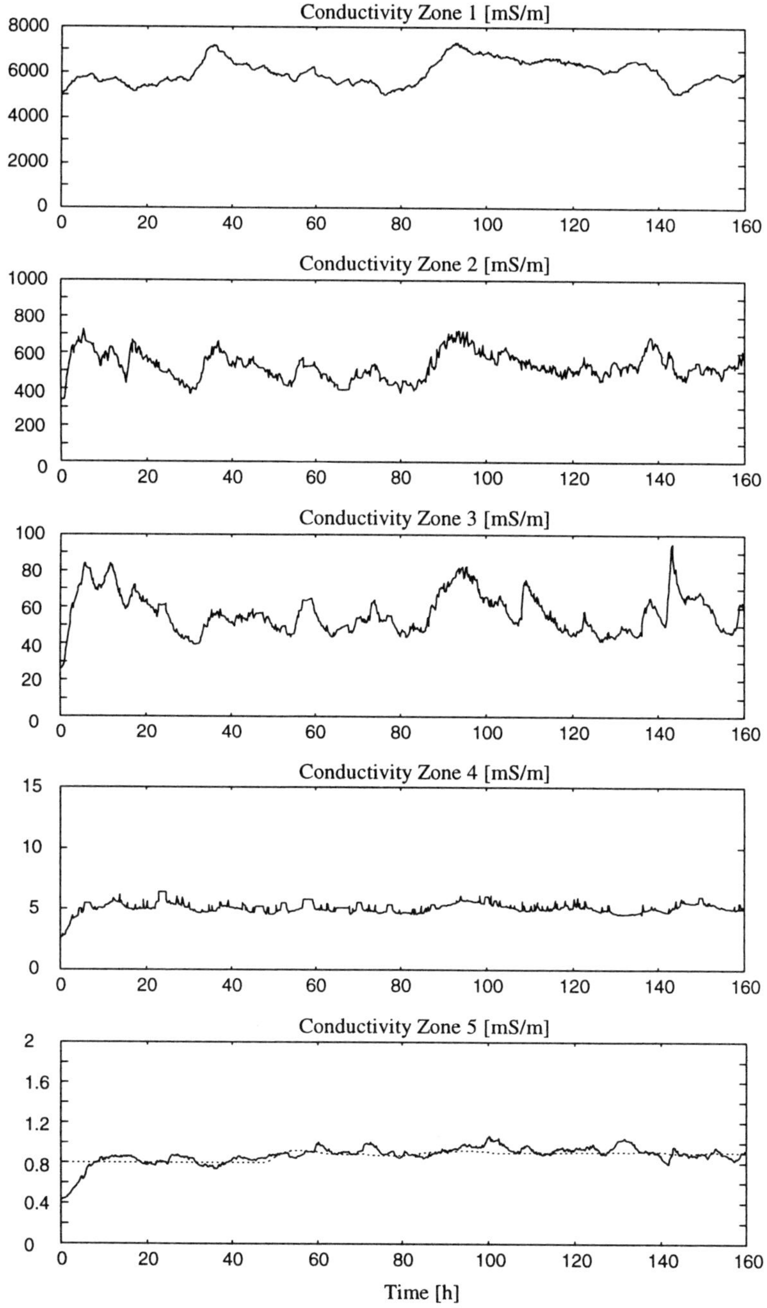

Fig. 6.19 Process outputs from control of the process, *dotted* - reference.

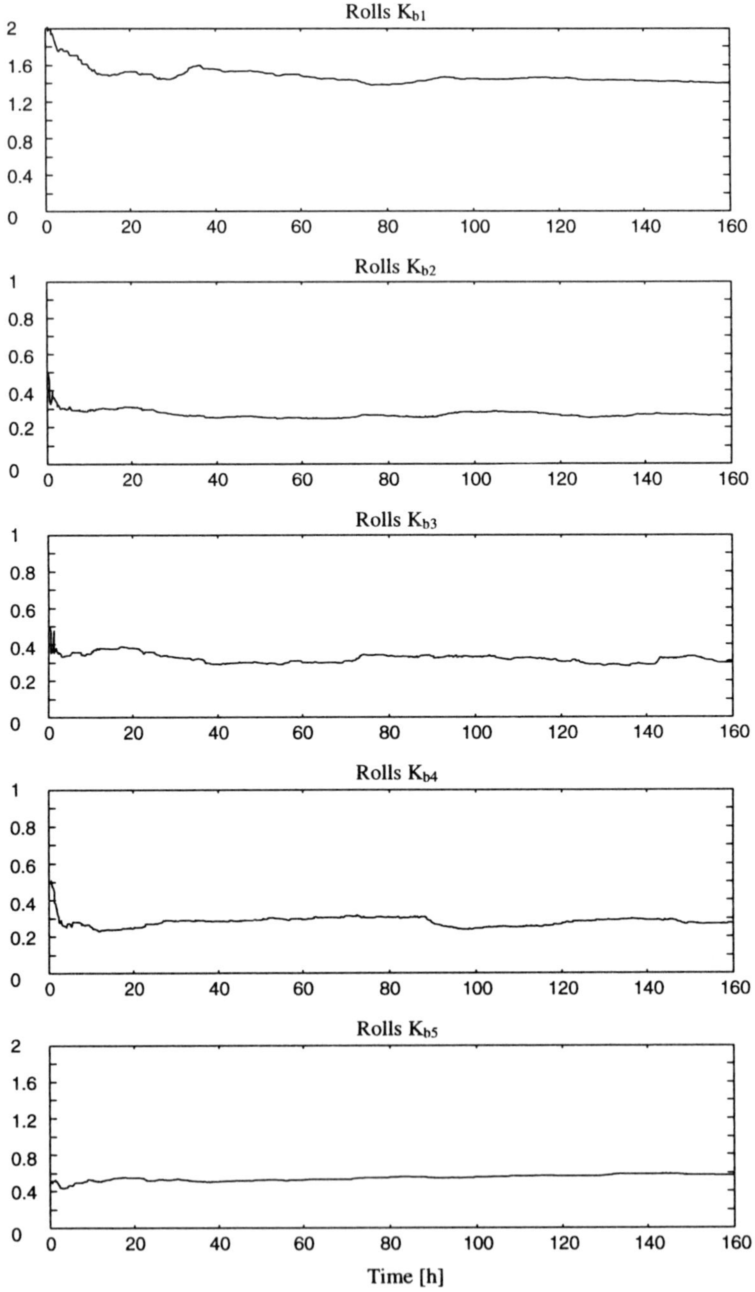

Fig. 6.20 Estimated parameters from control of the process, (K_{bi}).

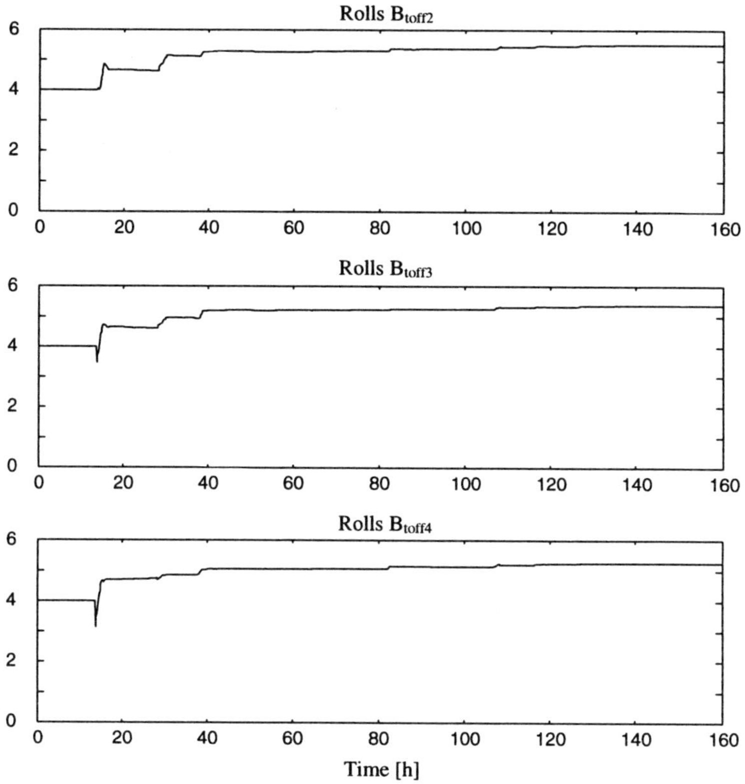

Fig. 6.21 Estimated parameters from control of the process, (B_{toffi}).

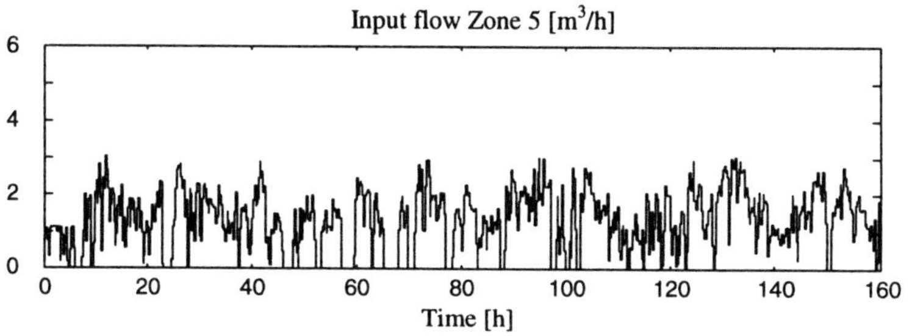

Fig. 6.22 Input signal from control of the process, ($u=F_{c1}$).

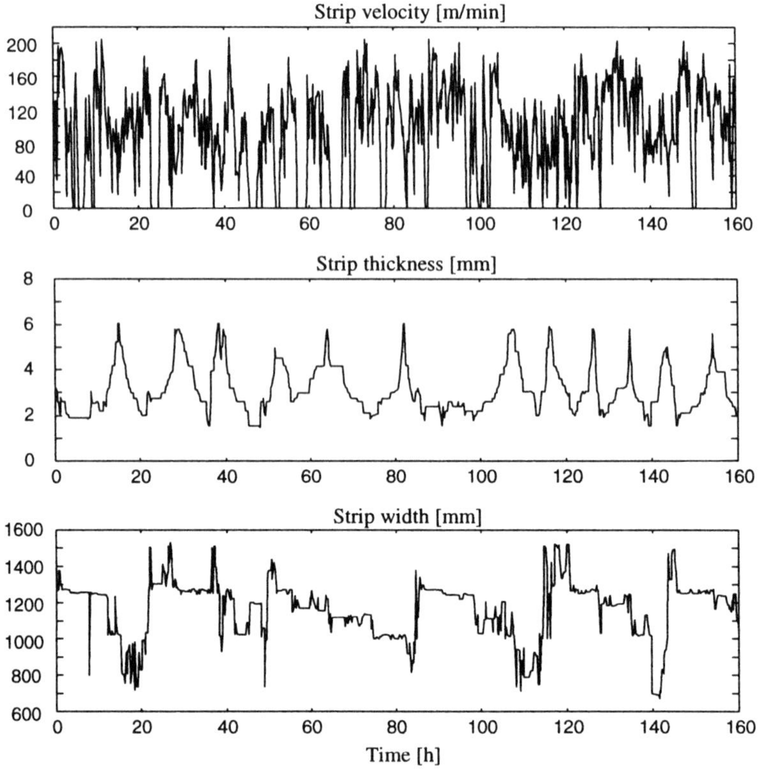

Fig. 6.23 Auxiliary inputs from control of the process, production variables.

6.6 Summary

In this Chapter we have discussed process control, based on a grey box model. The process is influenced by production variables, which are measurable but not possible to control. Furthermore, the process is influenced by disturbances which are not measurable. Two different control algorithms have been studied. One of the algorithms is a combined feedforward and feedback controller and is partly heuristic. The other algorithm is based on an optimal controller which is closely related to a general Model Predictive Controller.

The feedforward part of the combined controller compensates for variations in the production variables while the feedback part compensates for disturbances and simplifications in the control algorithm. The optimal controller is based on minimisation of a loss function for the future predicted behaviour of the process.

Simulations of the different control algorithms show that it is possible to compensate for variations in the production variables and disturbances acting on the process. By repeated simulations of the control system, the feasibility of the design parameters of the control algorithms are evaluated. The unknown parameters are also estimated and used by the controller.

The controllers are tested several times with the process. During tests with the combined controller we see that the output is mostly below the specified upper limit. The results of tests with the optimal controller show that the standard deviation of the output is better than using the combined controller. However, it is difficult to compare the experimental results because the parts subject to wear are replaced between the experiments.

From the results of estimation of the unknown parameters, we can see that the parts subject to wear used during the experiment with the optimal controller were more efficient than before. This is because the process operators were informed before the planned stops which parts were in bad condition and had to be replaced.

In Chapter 5 we have presented three different measurement sequences based on the normal feedforward controller. They are called sequence 1, sequence 2 and sequence 3. In table 6.1 statistical data are shown from these three sequences together with statistical data from the measurement sequences with the combined controller and the optimal controller. The later sequences are called sequence 4 and sequence 5 respectively.

Table 6.1 Statistical data

	Output signal [mS/m]			Control signal [m^3/h]
Sequence	Mean value	St. deviation	Var. interval	Mean value
No. 1	0.86	0.32	1.25	2.0
No. 2	0.86	0.38	1.67	1.9
No. 3	1.00	0.22	1.17	1.8
No. 4	0.77	0.12	0.40	1.3
No. 5	0.93	0.05	0.27	1.5

The mean value of the output signal is presented in the second column, the standard deviation presented in the third column and the variation interval presented in the fourth column. The mean value of the control signal is given in the fifth column.

The statistical values obtained from sequence 4 and sequence 5 are typical for ordinary process runs. Similar values have been achieved from other tests. From table 6.1, it is concluded that the combined controller reduces the standard deviation and the variation interval for the output signal from half to a third of the values achieved from the normal control of the process. It can also be noted that the control signal is reduced during control of the process with the combined controller and the optimal controller. This means that the new controllers are more efficient than the ordinary controller.

It can be concluded that the combined controller is partly heuristic and simpler to implement than the optimal controller. Nevertheless, we achieve a considerable improvement with the combined controller compared to the feedforward controller currently in use.

During the experiment, a PC-computer (80486 processor, 33 MHz) was used for estimation of the unknown parameters and to compute the control signal. All calculations were done with Matlab™ and we made new Matlab routines to communicate with the process interface cards. It takes about four seconds to compute the control signal with the combined controller and about 40 seconds to compute the control signal based on the optimal controller. This includes also the estimation of the unknown parameters. If a computer based on a Pentium processor was used, 200 MHz, the computation would be at least five times faster.

7 PROCESS SUPERVISION

7.1 Introduction

Supervision of industrial processes becomes more and more important together with of the increased level of automation in industry. Process performance monitoring and diagnosis technology are an emerging engineering area due to increased demands on product quality, low production cost, reduced cost for maintenance of the equipment and loss of production time due to production stops. Several of these factors are not only regulator problems but also problems on a higher process control level. The overwhelming purpose of process supervision is to improve the overall system performance.

Conventionally, supervision is carried out by checking values of some significant process variables. When a process is to be supervised, one way is to install more sensors and present more information from these sensors to the process operators. Since the number of signals increases, it will be difficult to analyse the information presented. Furthermore, many process variables cannot be measured by available sensors. Another way to solve the latter problem is to use process models, and with the aid of computers, estimate process parameters or by using statistical methods, which give a basis for decisions affecting the process.

An early overview of methods for failure detection in linear systems is given by Willsky (1976). The author comments on advantages and disadvantages concerning several techniques, defines several categories of failure detection system and collects them into groups. Isermann (1984) presents a survey paper focused on fault detection methodology based on modelling, parameter and state estimation.

A common type of faults originates from wear and the function of a part slowly detoriates due to long term wear. Mechanical systems, especially, have moving parts which are subject to wear. Many processes are also influenced by other faults apart from normal wear. These often occur abruptly and have to be detected very fast and be distinguished from wear. Typical abrupt failures are actuator failures, sensor failures and failures arising from a process unit malfunctioning or having stopped.

In this chapter, we focus on slowly varying changes of the efficiency of the process; abrupt faults are discussed in Chapter 8. Since, advanced process supervision is a relatively new subject, the terminology is not unambiguous. The Technical Committee of SAFEPROCESS has paid attention to this problem and several definitions are suggested, (Isermann and Ballé, 1996). A list of the most important definitions is given in Section 7.2. Methods for wear supervision are treated in Section 7.3. The methods presented are model based and varying parameters are estimated on-line. Here, process supervision is used for preventive maintenance and we give an economic view of how to form a strategy to change worn parts.

Concerning the rinsing process we wish to supervise the condition of the squeezer rolls, because they are subject to wear when the process is running. The efficiency rates of the rolls cannot be measured directly since there are no sensors to solve this measurement problem. The acidity of the rinse water depends however on the condition of the squeezer rolls. Consequently, to have a well running process, the squeezer rolls have to be in good condition.

We introduce a formalism to handle the efficiency of the process. The flow via the strip is related to a stationary point. This is done to make it possible to compare the efficiency rates of the different pairs of squeezer rolls. The compound influence of all pairs of squeezer rolls is also discussed. New guidelines are given as to when a pair of squeezer rolls should be replaced.

7.2 Terminology

To consider faults in the field of automatic control systematically is relatively new. The terminology has been developed by different people in the area, often separately from each other. This has resulted in terminology which is not uniform. However, the technique of maintenance is not new. It is applied to several disciplines in the space technology and the electronic industry, so there is a history to originate from.

The Technical Committee, SAFEPROCESS, has been working on the subject to find an acceptable frame of definitions in the field of supervision and fault detection. The discussion is still going on, so the proposed terminology is preliminary. A summary is presented by Isermann and Balle (1996); here it is used by permission of Professor Isermann. Additional terminology is given by Cuyvers et. alt (1990). The terminology is divided into: States and Signal, Functions, Models, and System Properties.

States and Signals

Fault: Unpermitted deviation of at least one characteristic property or variable of the system.

Failure: Permanent interruption of a system's ability to perform a required function under specific operating conditions.

Malfunction: Intermittent irregularity in fulfilment of a system's desired function.

Error: Deviation between a measured or computed value (of an output variable) and the true, specified or theoretically correct value.

Disturbance: An unknown (unmeasurable and uncontrolled) input acting on a system.

Perturbation: An input acting on a system which results in a temporary departure from steady state.

Residual: Fault indicator, based on model equations. The difference between model predictive and measured outputs.

Symptom: Change of an observable quantity from normal behaviour.

Functions

Fault detection: Determination of faults in a system and time of detection.

Fault diagnosis: Determination of kind, location and time of detection of a fault. Follows fault detection.

Fault isolation: Determination of kind, location and time of detection of a fault. Follows fault detection.

Fault identification: Determination of size and time-variant behaviour of a fault isolation.

Monitoring: A continuous real time task of determining the conditions of a physical system, by recording information recognising and indicating anomalies of the behaviour.

Supervision: Monitoring a physical system taking appropriate actions to maintain the operation in the case of faults.

Protection: Means by which potentially dangerous behaviour of the system is suppressed if possible, or means by which the consequences of a dangerous behaviour are avoided.

Models

Quantitative model: Use of static and dynamic relations among system variables and parameters in order to describe a system's behaviour in quantitative mathematical terms.

Qualitative model: Use of static and dynamic relations among system variables and parameters in order to describe a system's behaviour in qualitative terms such as causalities or `if - then´ rules.

Diagnostic model: A set of static and dynamic relations which link specific input variables (the symptoms) to specific output variables (the faults).

Analytical redundancy: Use of two or more, but not necessarily identical ways to determine a variable where one way uses a mathematical model in analytical form.

System Properties:

Reliability: Ability of a system to perform it's required function under stated conditions, within a given scope, during a given period of time. Specially useful when several systems are operating together and can be seen as the mean time between failure of some of the systems.
Measure: MTBF = Mean Time Between Failure.

Dependability: It is closely related to reliability, but is connected to a single system. It means Mean Time To Failure (MTTF).

Readability First Failure: It means Mean Time To First Failure (MTTFF).

Safety: Ability of a system not to cause danger for persons, equipment or environment.

Availability: Probability (A) that a system or equipment will operate satisfactorily and effectively at any point of time measure.

$$A = \frac{\text{MTBF}}{\text{MTBF} + \text{MTTR}}$$

MTTR is defined as Mean Time To Repair.

Dependability: A form of availability (D) that has the property of always being available when required. It is the degree to which an item is operable and capable of performing its required function at any randomly chosen time during its specified operating time, provided that the item is available at the start of that period.

$$D = \frac{\text{Time avaible}}{\text{Time availble } + \text{ Time requir}}$$

Integrity: The probability that a process will either perform its function correctly or will stop functioning in such a manner that no wrong outputs are generated.

Fault Avoidance: Any technique that is used to prevent faults.

Fault Tolerance: The ability of a system to perform a specified function even after the occurrence of a fault.

Fail Safe System: A system where fault tolerance is used to improve safety.

Redundancy: All additional source which is needed to make a system fault tolerant.

7.3 Supervision methods

7.3.1 On-line system

A key for monitoring wear in industrial processes is a feasible process model. Engineers use models for different purposes. Models for control describe the input - output behaviour of the process and are worked out with an identification procedure. Models for simulation relay on the description of the physical phenomena, in which non-linear differential equations are put together to form a system model. Those are two examples of task-oriented models; other examples would be models for process design and supervision.

Wearing is a slowly time-varying process and is a result of the efficiency of some process part decreasing. A general description of this situation is given in Fig. 7.1. The system involves the process and an estimator which is able to identify slowly time varying parameters. The estimates are used to form a basis for decision to affect the process.

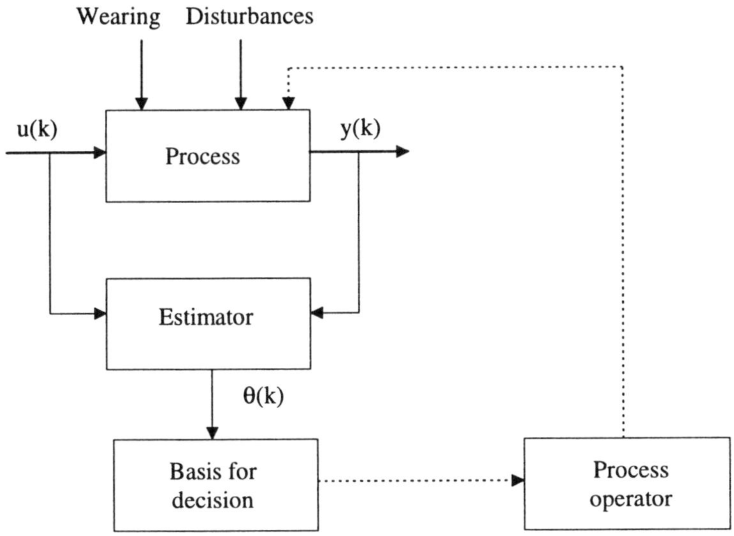

Fig. 7.1 Process subjected to wear.

The process is affected by the input signal u(k), wearing and disturbances. The estimator identifies the model parameters and where appropriate, estimates the process states. The estimation is based on a suitable process model, whose complexibility varies depending on the purpose of the supervision. It must be pointed out that the cost of the model is a part of the total supervising costs. Prerequisites for application of on-line parameter estimation method used for monitoring wearing are, (Isermann 1989):

- A process model which can describe the process behaviour sufficiently precisely.
- Powerful parameter estimation methods.
- Sufficient excitation of the process by appropriate input signals.
- Determination of maximum changes in process coefficients to achieve a well-running process. This means knowledge-based preventive maintenance.

7.3.2 Process models

The family of black-box models has been used in many applications as basis for on-line parameter estimation. These type of models can be considered as a linear description of the process, which is valid around an operating point. The models are basically linear regression models:

$$y(k) = \phi(k)^T \theta(k) + v(k) \tag{7.1}$$

where $\phi(k)$ is the regressor vector, $\theta(k)$ the parameter vector and v(k) is the disturbances acting on the system. The regressor is formulated as:

$$\phi(k) = \left[-y(k-1) \; \ldots \; -y(k-na) \;\; u(k-1) \ldots \; u(k-nb)\right]^T \tag{7.2}$$

A common black-box model is the ARMAX (Auto Regressive Moving Average exogenous input). The equations are written in a difference form:

$$A(z^{-1})y(k) = z^{-d}B(z^{-1})u(k) + C(z^{-1})v(k) \tag{7.3}$$

where d is a time delay and A, B and C are polynomials and v(k) is white noise.

$$A(z^{-1}) = 1 + a_1 z^{-1} + \ldots + a_{na} z^{-na} \tag{7.4}$$

$$B(z^{-1}) = b_0 + b_1 z^{-1} + \ldots + b_{nb} z^{-nb} \tag{7.5}$$

$$C(z^{-1}) = 1 + c_1 z^{-1} + \ldots + c_{nc} z^{-nc} \tag{7.6}$$

with the parameter vector:

$$\theta(k) = \begin{bmatrix} a_1 \ldots a_{na} & b_0 \ldots b_{nb} & c_1 \ldots c_{nc} \end{bmatrix}^T \tag{7.7}$$

Another way of describing a black box model is by a transfer function which separates the deterministic and stochastic part. There are mainly two separate transfer function models: the output error model and the Box-Jenkins model:

$$y(k) = \frac{B(z^{-1})}{F(z^{-1})} u(k) + v(k) \tag{7.8}$$

$$y(k) = \frac{B(z^{-1})}{F(z^{-1})} u(k) + \frac{C(z^{-1})}{D(z^{-1})} v(k) \tag{7.9}$$

For the model described by equation (7.8), we assume that the disturbances act directly on the process output, but by the Box Jenkins model the disturbances are modelled by a separate transfer function.

A general description which includes both difference equations and transfer functions is:

$$A(z^{-1}) y(k) = \frac{B(z^{-1})}{F(z^{-1})} u(k) + \frac{C(z^{-1})}{D(z^{-1})} v(k) \tag{7.10}$$

where the model parameters are formulated by:

$$\theta(k) = \begin{bmatrix} a_1 \ldots a_{na} & b_0 \ldots b_{nb} & c_1 \ldots c_{nc} & d_1 \ldots d_{nd} & f_1 \ldots f_{nf} \end{bmatrix}^T \tag{7.11}$$

Generally, a process can be described by a higher order differential equation. It is possible to transform the model into a finite number of first order difference equations. For a linear model, we introduce a discrete state space vector $x(k) \in R^n$, input vector $u(k) \in R^r$ and output vector $y(k) \in R^p$.

$$\begin{cases} x(k+1) = \Phi(k)x(k) + \Gamma(k)u(k) + v(k) \\ y(k) = Cx(k) + w(k) \end{cases} \tag{7.12}$$

where v(k) and w(k) are independent white noise with covariance matrices R_1 and R_2. The matrices Φ, Γ and C correspond to the parameter vector θ given by equation (7.11).

Normally, a linear model behaves satisfactorily in the neighbourhood of a defined operating point. This means that it is a difficult to separate the influence of being away from the operating point and the influence originating from wear. One possibility can be to analyse the output signal. The signal can often be separate in a lower frequency and a higher frequency part, and by statistical methods produce more information.

Another method is to construct a model based on physical insight. Since the process knowledge is incomplete often, the grey box modelling technique is a practical way of achieving a model based on parameters which have a physical meaning. In this way it is possible to let the model consist of parameters, which describe the wearing process explicitly.

Such a model is generally described by:

$$\begin{cases} x(k+1) = F[x(k), u(k), \theta(k)] + v(k) \\ y(k) = G[x(k), \theta(k)] + w(k) \end{cases} \tag{7.13}$$

where F and G are non-linear functions, and $\theta(k)$ is a slowly varying parameter vector.

7.3.3 On-line estimation

When we are dealing with time varying processes, the process parameters will change. This means that the values of the parameters estimated will not be valid after a certain time. Therefore, it may be necessary to compute process parameters on-line. The measured input-output signals are used to update the variables as soon as the signals are available.

For wear supervision, the updated parameters are to form a basis to replace worn parts. We must be aware of other sources that may change the parameters, for instance, changing operation point and disturbances acting on the process.

There are many textbooks dealing with the subject of on-line estimation. The history is long. The first method of forming an optimal estimate from noisy data is the least square method by C. F. Gauss, (1795). Modern textbooks within recursive estimation have been written by Söderström and Stoica (1989), Ljung (1987), Johansson (1993) and Goodwin and Payne (1977). A special textbook for recursive identification has been written by Ljung and Söderström (1983). Adaptive control is closely related to recursive estimation; this subject is treated by Åström and Wittenmark (1995) and Isermann et. alt (1992). Several on-line estimation methods originate from off-line methods like the least square method.

Least Square Recursive Estimation

This algorithm is often used in the field of adaptive systems. First, we introduce the forgetting factor, $\lambda<1$. The factor is used to reduce the influence of old data on the estimate. The influence of the latest measurement is weighted by the unity factor but the previous n:th measurement is affected by λ^n. Introduce the parameter estimate $\hat{\theta}(k)$, the regressor $\phi(k)$, the prediction error $\varepsilon(k)$ and the covariance matrix P(k). This gives a recursive algorithm:

$$\hat{\theta}(k) = \hat{\theta}(k-1) + P(k)\phi(k)\omega(k) \qquad \hat{\theta}(0) = \theta_0 \tag{7.14}$$

$$\omega(k) = y(k) - \phi(k)^T \hat{\theta}(k-1) \tag{7.15}$$

$$P(k) = \frac{1}{\lambda}\left[P(k-1) - \frac{P(k-1)\phi(k)\phi(k)^T P(k-1)}{\lambda + \phi(k)^T P(k-1)\phi(k)}\right] \tag{7.16}$$

The value of the forgetting factor λ determines the sensitivity to changes. The choice of the value is a trade-off between the possibility to track fast variations of the parameter vector $\theta(k)$, and reject the influence of disturbances on the estimate. A common choice of λ is between 0.970 and 0.995.

Kalman Filter Estimation

Introduce the estimated parameters as a state vector and model the state equation as a random walk process, equation (7.17). The output is achieved from the linear regression model, given by equation (7.18).

$$\theta(k+1) = \theta(k) + v(k) \tag{7.17}$$

$$y(k) = \phi(k)^T \theta(k) + w(k) \tag{7.18}$$

The covariance matrices for the disturbances v(k) and w(k) are R_1 and R_2 Applying the Kalman Filter to the system results in the following recursive algorithm:

$$\hat{\theta}(k) = \hat{\theta}(k-1) + K(k)\omega(k) \tag{7.19}$$

$$K(k) = \frac{P(k-1)\phi(k)}{R_2 + \phi(k)^T P(k-1)\phi(k)} \tag{7.20}$$

$$\omega(k) = y(k) - \phi(k)^T \hat{\theta}(k-1) \tag{7.21}$$

$$P(k) = P(k-1) - \frac{P(k-1)\phi(k)\phi(k)^T P(k-1)}{R_2 + \phi(k)^T P(k-1)\phi(k)} + R_1 \tag{7.22}$$

The estimation is tuned by affecting the covariance matrices R_1 and R_2. Since R_2 has a physical meaning of measurement noise, it is preferable to let R_2 be constant and change R_1. A large vale of R_1 gives fast tracking possibility but high sensitivity to disturbances. An important difference between the Least Square Recursive and Kalman Filter methods is:

- The estimated covariance matrix, P(k), for the Kalman Filter does not go to zero when k goes to infinity. This means that the Kalman gain K(k) will not approach zero.
- The covariance matrix, P(k), has different dynamics for the two methods when $\phi(k)=0$. For the Least Square method P(k) increases exponentially but for the Kalman Filter method the growth rate is linear.

Augmented Extended Kalman Filter

In order to solve the problem of simultaneous estimation of both states and unknown parameters recursively, the state vector is augmented with the unknown parameters and the Extended Kalman Filter is used to solve the resulting non-linear filtering problem, (Balchen et al, 1992). This strategy is often referred to as the Augmented Extended Kalman Filter, (AEKF). However, the method has a drawback, which means that the estimates may diverge. Further details on the practical implementation can be found in Maybeck (1979 and 1982). We form a non-linear discrete state space model:

$$x(k+1) = F[x(k), u(k), \theta(k)] + v(k) \tag{7.23}$$

$$y(k) = G[x(k)] + w(k) \tag{7.24}$$

The matrix F involves unknown time varying parameters. They are formed as a state vector and modelled as a random walk process:

$$\theta(k+1) = \theta(k) + v_\theta(k) \tag{7.25}$$

The augmented state vector $\xi(k)$ and the new model are given as:

$$\xi(k) = \begin{bmatrix} x(k) \\ \theta(k) \end{bmatrix} \tag{7.26}$$

$$\xi(k+1) = \overline{F}[\xi(k), u(k)] + \overline{v}(k) \tag{7.27}$$

$$y(k) = \overline{G}[\xi(k)] + \overline{w}(k) \tag{7.28}$$

The model given by equations (7.27) and (7.28) is affected by white noise with the covariance matrices $\overline{R}_1$ and $\overline{R}_2$.

7.3.4 Preventive maintenance

System monitoring can be used to present actual states and trends of the process, giving advanced information when the process is going to malfunction. This can be used in planning the maintenance of the system and allowing parts to be ordered to arrive in time and be replaced without delay. This adds to the advantages of generally lower damage levels and shortens the time needed to maintain the system.

Conventionally, maintenance schedules are based on fixed time intervals. This time is often set by intuition or experience. It takes no account of the working condition of the process and maintenance will occur more often than needed to preserve the reliability of those parts in poor condition. As wear in the system increases, the produced products may not be rejected but give a general decrease in final quality. If the cause of the quality problem can be monitored, the trend towards an unacceptable level can be visualised and the problem can be taken care of by maintenance. When we are using a process model, which includes parameters describing the degree of wear, it is possible to compensate this by adjusting the control signal. This means, the wear is compensated by an adaptive controller, which maintains the final quality.

A system with a controller compensated for wear involves extra costs, since an increased control signal results in higher expense. It is also possible that a higher degree of wear will cost more to renovate than a less worn part. On the other hand, the total maintenance time will decrease, since the worn parts can be used for a longer time. Based on this discussion, it is possible to formulate a cost function for preventive maintenance:

$$J = \sum_{k=1}^{N} \|u(k)\|^2_{\Lambda_u} + \|\tau(N)\|^2_{\Lambda_\tau} + \|c(N)\|^2_{\Lambda_c} \qquad (7.29)$$

N	The number of samples between the maintenance stops.
Λ_u, Λ_τ and Λ_c	General weighting matrices.
$\|\ \|^2$	The quadratic norm.
$\tau(N)$	Costs due to maintenance stops
$c(N)$	Costs for spare parts.
$u(k)$	Control signal.

❐ **Remark.** When the number of samples between the stops increases, the costs for production loss decrease and normally, since the wearing parts can be used for a longer time the cost for spare parts will also decrease.

❐ **Remark.** A special case of the cost function is for the situation when an increase in the control signal is not accomplished with higher costs. In this situation the optimal supervision strategy is to run the process as long as the control signal can compensate for wearing.

❐ **Remark.** It should be pointed out that the costs for developing the process model are also a part of the total supervision costs.

7.4 Application of supervision

7.4.1 System design

We make a similar approach as in Section 7.3, for supervision of the squeezer rolls, Sohlberg (1992). In Chapter 4 we have developed a model of the rinsing process based on physical laws. This model reflects the internal behaviour of the process. The estimation of the unknown parameters is made with the Augmented Extended Kalman Filter, see Chapter 5. This type of filter is appropriate since the values of the estimates changes slowly compared to the process dynamics.

The on-line supervision system is shown in Fig. 7.1. The parameter estimation is made in the same way as in Chapter 5. We use the same initial values and the same values of the covariance matrices. The process operator shown in Fig. 7.1 belongs to the supervision system since the system is an advisory system; the final decision on which squeezer rolls should be replaced is taken by the process operator. The reason why the system is an advisory system is that the process operator should take an active part in maintenance of the process.

7.4.2 Decision basis for replacing worn parts

Stationary reduction

To achieve a well-rinsed steel strip, the conductivity in the rinse tank must be lower than 1.0 [mS/m]. This value is based on experience of running the rinsing process. From Chapter 3 and 4, we know that the concentration in rinse tank 5 depends on the flow via the strip and the flow of clean water fed into rinse tank 5.

The unknown parameters in the process model are estimated on-line. These parameters are used to compute the flow via the steel strip, which also depends on the production variables. The relation between the flow via the strip and the production variables is given by equation (3.12). It is seen that the flow via the strip consists of two parts and is non-linear depending on the production variables.

The flows within the process are described in Chapter 3 and are also shown in Fig. 7.2. In the figure, the flow via the strip is named F_{b1} to F_{b5}; the main flow is named F_{m1} to F_{m5}. The controlled flow of clean water is named F_{c1} and F_{c2}. The concentrations in the rinse tanks are named C_1 to C_5.

Fig. 7.2 Flows and concentrations.

Suppose that the process is in a stationary state. Equation (3.26) gives the following for rinse tank 5:

$$F_{b5} \cdot C_5 - F_{m5} \cdot C_5 = 0 \qquad (7.30)$$

From equation (3.31) we have:

$$F_{m5} = F_{c1} + F_{b5} \tag{7.31}$$

The reduction in concentration at a stationary point between rinse tank 4 and rise tank 5 is defined by:

$$\eta = \frac{C_4}{C_5} \tag{7.32}$$

Equation (7.30) gives:

$$\frac{C_4}{C_5} = \frac{F_{m5}}{F_{b5}} \tag{7.33}$$

Equation (7.31) - (7.32) give that the reduction between rinse tank 4 and rinse tank 5 is:

$$\eta_5 = 1 + \frac{F_{c1}}{F_{b5}} \tag{7.34}$$

For rinse tank 4, we obtain in the same way from equation (3.25):

$$F_{m5} \cdot C_5 - F_{b5} \cdot C_4 + F_{b4} \cdot C_3 - F_{m4} \cdot C_4 = 0 \tag{7.35}$$

Equation (7.35) can be rewritten in the following form:

$$\frac{C_3}{C_4} = \frac{F_{m4}}{F_{b4}} + \frac{F_{b5}}{F_{b4}} - \frac{F_{m5}}{F_{b4}} \cdot \frac{C_5}{C_4} \tag{7.36}$$

Equation (7.33) and (7.36) give the reduction between rinse tank 3 and rinse tank 4:

$$\frac{C_3}{C_4} = \frac{F_{m4}}{F_{b4}} \tag{7.37}$$

From equation (3.30) we get:

$$F_{m4} = F_{m5} + F_{b4} - F_{b5} \qquad (7.38)$$

Equation (7.31) and (7.38) give:

$$F_{m4} = F_{c1} + F_{b4} \qquad (7.39)$$

From equation (7.37) and equation (7.39), the reduction between rinse tank 3 and rinse tank 4 is given by:

$$\eta_4 = 1 + \frac{F_{c1}}{F_{b4}} \qquad (7.40)$$

In the same way as above, the reduction between rinse tank 2 and rinse tank 3 is given by:

$$\eta_3 = 1 + \frac{F_{c1}}{F_{b3}} \qquad (7.41)$$

When formulating the reduction between rinse tank 1 and rinse tank 2 we had to include the flow F_{c2}, which is the controlled flow of clean water fed directly into rinse tank 2. This gives the reduction between tank 1 and tank 2 as:

$$\eta_2 = 1 + \frac{F_{c1} + F_{c2}}{F_{b2}} \qquad (7.42)$$

It should be noted that the reduction in concentration between the rinse tanks is equal to the reduction in the conductivity between the tanks. This is given by the fact that the conductivity is proportional to the concentration of hydrochloric acid, see equation (3.37). The reduction in concentration or conductivity between the rinse tanks can be summed up by the following equations:

$$\eta_i = \frac{c_{i-1}}{c_i} \qquad (i=2...5) \qquad (7.43)$$

$$\eta_i = 1 + \frac{F_{c1} + F_{c2}}{F_{bi}} \qquad (i=2) \qquad (7.44)$$

$$\eta_i = 1 + \frac{F_{c1}}{F_{bi}} \qquad (i=3...5) \qquad (7.45)$$

It must be pointed out that the equations (7.43) - (7.45) are related to a representative stationary point.

Compound influence of the squeezer rolls

As discussed earlier, the conductivity in tank 5 must be lower than 1.0 [mS/m]. The conductivity in tank 5 is related to the conductivity in tank 1 by the reduction between the tanks. Let y_1 be the conductivity in tank 1 and y_5 be the conductivity in tank 5. The definition of reduction gives:

$$y_1 = \eta_2 \cdot \eta_3 \cdot \eta_4 \cdot \eta_5 \cdot y_5 \qquad (7.46)$$

When $Con_5 \le 1.0$, equation (7.46) gives:

$$\eta_2 \cdot \eta_3 \cdot \eta_4 \cdot \eta_5 \ge y_1 \qquad (7.47)$$

The natural logarithm of equation (7.47) gives:

$$\ln(\eta_2) + \ln(\eta_2) + \ln(\eta_2) + \ln(\eta_2) \ge \ln(y_1) \qquad (7.48)$$

Equation (7.48) is rewritten in following compact form:

$$\sum_{i=2}^{5} \ln(\eta_i) \ge \ln(y_1) \qquad (7.49)$$

Equation (7.49) gives the necessary condition to achieve a conductivity value in rinse tank 5 less than 1.0 [mS/m]. Note, the condition can be achieved in several ways. This means that a bad pair of squeezer rolls can be compensated for by good rolls, and we still achieve sufficiently clean rinse water in tank 5.

Choice of stationary point

From equation (7.44) and (7.45) we have the reduction in the conductivity value, which is related to a combination of process inputs. To compare the efficiency rates of the squeezer rolls we have related the reduction to a stationary point.

The strip velocity and the strip width vary around a mean value. These variables are distributed with roughly the same time above as below the mean value. It is, therefore possible to conceive a stationary point as a mean value of these variables. However, the strip thickness varies in a different way. The periods when thin steel strips are rinsed are longer than the periods when thick steel strips are rinsed.

In Chapter 5, we have discussed estimation of the unknown parameters. One of the estimated parameters is related to the flow beside the strip. The strip thickness has to exceed this parameter before the model is influenced by the flow beside the strip. The lowest value of this estimated parameter is about 3.0 [mm]. The maximum value of the strip thickness is about 6.0 [mm]. Consequently, within the interval 3 - 6 [mm], the strip thickness influences the flow via the strip.

If the stationary point of the strip thickness is chosen as the mean value of the interval, the strip thickness has some influence on the model, which means we obtain a value of 4.1 [mm]. This value is taken as a representative value for a suitable stationary point. The three measurement sequences which were discussed in Chapter 5, each give about the same values for the stationary point. This yields a suitable stationary point given in table 7.1.

Table 7.1 Stationary point of the process.

Process variable	Stationary point	Unit
Strip velocity	90	[m/min.]
Strip width	1170	[mm]
Strip thickness	4.1	[mm]
Flow into tank 1	1.8	[m^3/h]
Flow into tank 2	0.1	[m^3/h]

Profile of the conductivity within the process

We introduced the reduction in concentration between the rinse tanks, which is equivalent to the reduction in the conductivity value. The reduction is related to a stationary point, which is representative of the process. The reduction is computed from equation (7.44) and (7.45):

$$\eta_i = 1 + \frac{F_{c1} + F_{c2}}{F_{bi}} \qquad (i=2) \qquad (7.50)$$

$$\eta_i = 1 + \frac{F_{c1}}{F_{bi}} \qquad (i=3...5) \qquad (7.51)$$

where Fb_i is computed from equation (7.52). See also Chapter 4..

$$F_{bi} = K_{bi} \cdot B_v \cdot B_b + K_{ti} \cdot B_v \cdot (B_t - B_{toffi}) \cdot \alpha_i \qquad (7.52)$$

$$\begin{cases} \alpha_i = 0 & \text{if } B_t \le B_{toffi} \\ \alpha_i = 1 & \text{if } B_t \ge B_{toffi} \end{cases} \qquad (7.53)$$

In equation (7.52), K_{bi} and B_{toffi} are the unknown parameters estimated on-line. K_{ti} is chosen as a constant value. The unknown parameters are estimated for three measurement sequences; sequence 1, sequence 2 and sequence 3, see Chapter 5.

The estimated parameters will be used to achieve a basis to decide which pair of squeezer rolls is to be replaced. We use the end value of the estimates from the sequences. With the stationary values of the inputs we compute the reduction from equations (7.50), (7.51) and (7.52). The reductions for the three measurement sequences are shown in table 7.2. The pairs of squeezer rolls which are replaced between sequence 1 and sequence 2 are denoted with the symbol: ☞.

Table 7.2 Reduction between the rinse tanks.

Squeezer rolls	Sequence 1		Sequence 2	Sequence 3
2	5.52	☞	19.86	19.86
3	9.37		6.32	6.04
4	12.91	☞	8.86	7.55
5	5.62		4.27	3.46

❒ **Comments on the results in table 7.2**

When the efficiency of the squeezer rolls is good, the reduction is high. From table 7.2 it is shown that after sequence 1, the squeezer rolls 2 are in worst condition. After the replace of the squeezer rolls 2, it is shown that the efficiency of this pair is considerably increased.

The pair of squeezer rolls 3 was not replaced during the planned stop. It can be noted that the condition of this pair is worse after sequence 2 and sequence 3.

The pair of squeezer rolls 4 was replaced during the planned stop. This pair was in best condition compared to the other pairs of squeezer rolls. It can be noted that the efficiency of this pair is decreased after the replace of the squeezer rolls.

The pair of squeezer rolls 5 was not replaced during the planned stop. The condition of this pair is roughly the same as the condition of squeezer rolls 2 after sequence 1. Note that squeezer rolls 2 were replaced, but not squeezer rolls 5. It should also be noted that the efficiency of this pair was decreased after sequence 2 and sequence 3.

To sum up: by computing the reduction of the concentrations between two adjacent rinse tanks, we have a measure of the efficiency of the squeezer rolls. By comparing the reduction, we have information about which pair of squeezer rolls is in best and which in worst condition. This gives guidance on which pair of squeezer rolls should be replaced during the next planned stop for maintenance.

Since the reduction is dependent on the stationary point, we have analysed the sensitivity of the reduction. This is done by computing the reduction when the stationary point is changed by 10%. The results are that a change of strip velocity, strip width and flow into tank 5 will affect the reduction by less than 10%. A change of strip thickness will affect the reduction by about 15%. When the stationary point is changed, the relationships between the reductions in table 7.2 are roughly preserved.

Equation (7.49) means that the sum of the natural logarithm of the reduction gives the conductivity value in tank 5. Consequently, when we know the conductivity in rinse tank 1, it is possible to decide if the compound reduction is large enough. In table 7.3, the natural logarithm of the reductions and the sum of the reductions are shown.

Table 7.3 Natural logarithms of the reductions.

Squeezer rolls	Data sequence 1		Data sequence 2	Data sequence 3
2	1.71	☞	2.99	2.99
3	2.24		1.84	1.80
4	2.56	☞	2.18	2.02
5	1.73		1.45	1.24
Sum	8.24		8.46	8.05

❐ **Comments on table 7.3.**

The sum of the reductions is increased between sequence 1 and sequence 2. This means that the compound condition of the squeezer rolls is improved after the change of squeezer rolls. Consequently, the effect of the change of the rolls was positive for the compound efficiency of the process. However, the total effect was slight, due to the bad decision to change squeezer rolls 4. The compound efficiency is reduced after sequence 3. This is a result of the fact that the condition of the squeezer rolls has deteriorated during the process operation.

Based on the conductivity value in rinse tank 1, it is possible to compute the corresponding stationary values of the conductivity value in the other rinse tanks. In Fig. 7.3, the stationary conductivity value in the other rinse tanks is shown.

From Fig. 7.3 it is shown that the conductivity value in rinse tank 2 is considerably decreased after the change of squeezer rolls 2. The improvement is lost in the later stages of the rinsing process because squeezer rolls 4 and 5 have deteriorated. In a comparison between the efficiency of squeezer rolls after sequence 2 and sequence 3 it is shown that squeezer rolls 4 and 5 deteriorated most. From Fig. 7.3 it is also shown that the degree of cleanness in rinse tank 5 can be achieved in several ways. It is shown that a bad pair of squeezer rolls can be compensated for by a good pair of squeezer rolls.

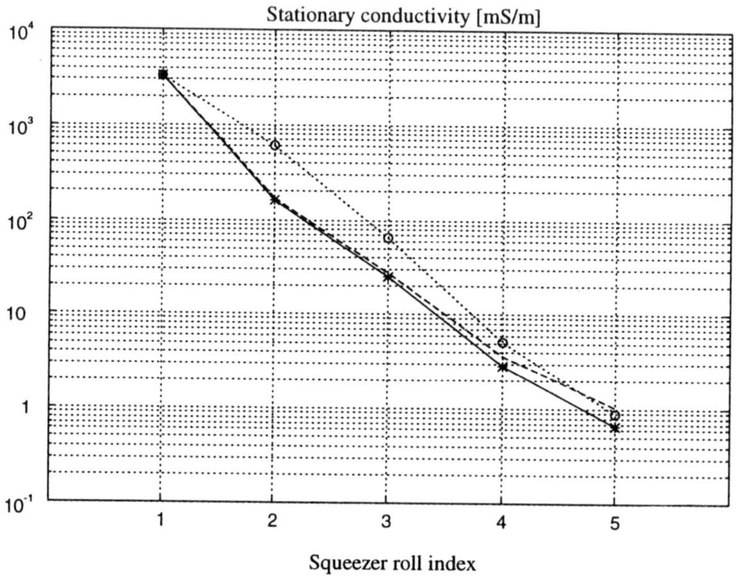

Fig. 7.3 Stationary conductivity profile for sequence 1, 2 and 3, *dotted* - sequence 1, *solid* - sequence 2 and *dotted/solid* - sequence 3.

7.4.3 Useful life of the squeezer rolls

The efficiency of the squeezer rolls detoriates when the process is running: the surface of the squeezer rolls becomes rougher and the flow via the strip increases. Furthermore, the quality of the squeezer rolls is individual, and the hardness of the rubber varies from squeezer roll to squeezer roll.

In Chapter 6, the process is controlled. It was concluded that the controllers are able to control the process so the conductivity in rinse tank 5 is below the limit. The conductivity value in rinse tank 1 is related to the conductivity value in rinse tank 5 by equation (7.46). Let the reduction, η_i i=2..5, between adjacent rinse tanks be equal. Equation (7.48) gives the same reduction from tank 1 to tank 5 as:

$$\bar{\eta} = \sqrt[4]{\frac{y_1}{y_5}} \qquad (7.54)$$

Let the flow into rinse tank 2 be equal to zero, $F_{c2}=0$. This is not a serious restriction because the controllers used in Chapter 4 do not use this flow of clean water as an input signal. Assuming that the reduction between the rinse tanks is equal, we have the flow via the strip from equation (7.51):

$$F_b = \frac{F_{c1}}{\bar{\eta} - 1} \qquad (7.55)$$

There is an upper limit when the controller cannot compensate for worn out squeezer rolls, because the production of clean water is limited to 3.0 [m^3/h]. However, the flow of clean water into rinse tank 5 is limited to 6.0 [m^3/h]. This is because the production of clean water is stored in a reservoir of 10.0 [m^3] before the clean water is fed into rinse tank 5. The reservoir is used as a buffer. The volume of the water in the reservoir increases when the flow from the reservoir is less than 3.0 [m^3/h] and decreases when the flow from the reservoir is more than 3.0 [m^3/h]. When the reservoir is full, the flow into tank 5 can be 6.0 [m^3/h] for 3.3 [h] without the reservoir becoming empty.

The mean flow into tank 5 should not exceed 3.0 [m^3/h], because this is the limit of the production of clean water. In Fig. 4.9, we can see that the amplitude of the flow into zone 5, which is high, is comparatively short. Therefore, if we let the stationary point of the flow into tank 5 be 3.0 [m^3/h], it is reasonable to believe that the reservoir is large enough to take care of the variations in the control signal.

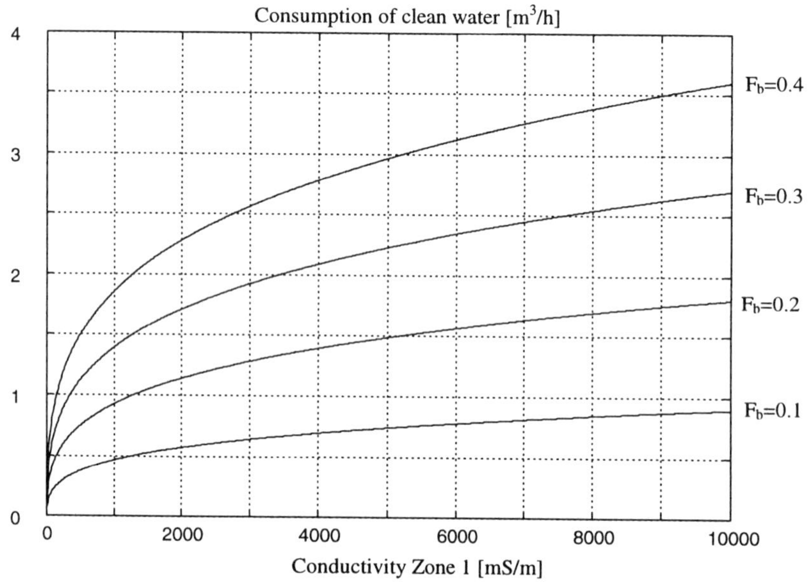

Fig. 7.4 Consumption of clean water.

By eliminating $\bar{\eta}$ in equation (7.54) and (7.55) we obtain equation (7.56).

$$F_{c1} = F_b \cdot \left\{ \sqrt[4]{\frac{y_1}{y_5}} - 1 \right\} \tag{7.56}$$

Equation (7.56) gives a relation between the consumption of clean water, conductivity in tank 1 and the flow via the strip. This kind relation is shown in Fig. 7.4.

In Fig. 7.4, we have four different curves depending on the estimated flow via the strip. If the conductivity value in tank 1 is 6000 [mS/m] and the flow via the strip is estimated to be 0.2 [m^3/h], we read from the diagram a consumption of clean water of about 1.6 [m^3/h].

Since the production of clean water is limited to 3.0 [m^3/h], there is a limit when the main flow cannot compensate for the flow via the strip. Equation (7.56) gives the limitation of the flow via the strip as:

$$F_b \leq \frac{3.0}{\sqrt[4]{\frac{y_1}{y_2}} - 1} \tag{7.57}$$

The limit of the flow via the strip is shown in Fig. 7.5. The conductivity in tank 1 is between 3000 and 9000 [mS/m]. When the flow after a pair of squeezer rolls is above the curve in Fig. 7.5 the squeezer rolls should be replaced.

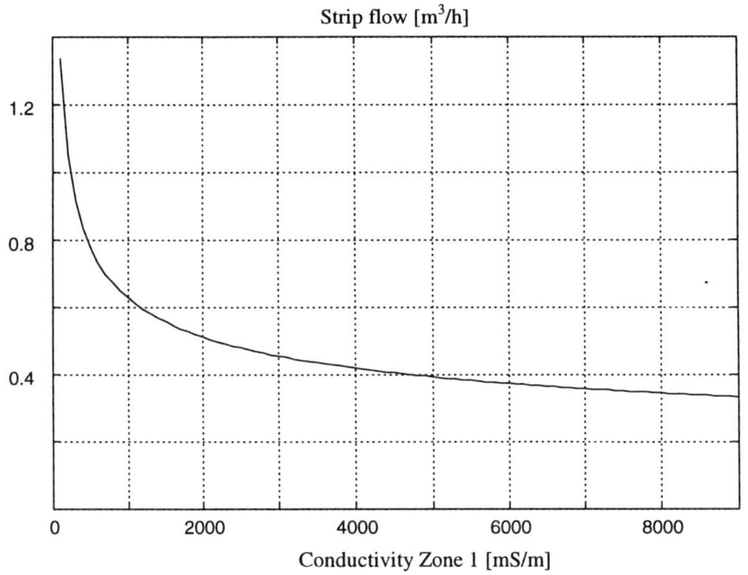

Fig. 7.5 Limitation of the flow via the strip.

The flow via the strip is computed for the two experiments given in Chapter 6. The flow at the stationary point is based on the estimated parameters obtained from the real experiment with the process. The results are presented in table 7.4. Note that table 7.4 only gives a measure of the efficiency of the squeezer rolls which were in use during the experiments and is not a comparison between methods.

During the experiment with the combined and the optimal controller, the conductivity value in tank 1 was about 6000 [mS/m]. This value gives a limit of the flow via the strip as 0.37 [m^3/h], see Fig. 7.5.

Table 7.4 Flow via the strip at the stationary point, [m^3/h].

Squeezer rolls	Combined controller	Optimal controller
2	0.09	0.12
3	0.11	0.16
4	1.61	0.12
5	0.11	0.26

❒ **Comments on table 7.4.**

It is concluded that during the experiment with the combined controller, the pairs of squeezer rolls 2, 3 and 5 were in good condition. However, the pair of squeezer rolls 4 has passed the limit, 0.37 [m^3/h], when the squeezer rolls were planned be replaced. During the experiment with the optimal controller no pair of squeezer rolls has passed the limit for change, but it can be noticed that the pair of squeezer rolls 5 was in worst condition.

The decision on which pair of squeezer rolls should be replaced is based on the estimated parameters. It is interesting to know how much an estimated parameter can be changed before the rolls become inefficient.

From equation (7.52), we have the relation between the flow via the strip and the estimated parameters. At the stationary point, the flow via the strip depends on the two parameters, K_b and B_{toff}. In Fig. 7.6, the flow via the strip is shown as a function of K_b, at three different values of B_{toff}. From equation (7.56), we have a relation between the flow via the strip, consumption of clean water and conductivity in tank 1. This relation is presented in Fig. 7.7.

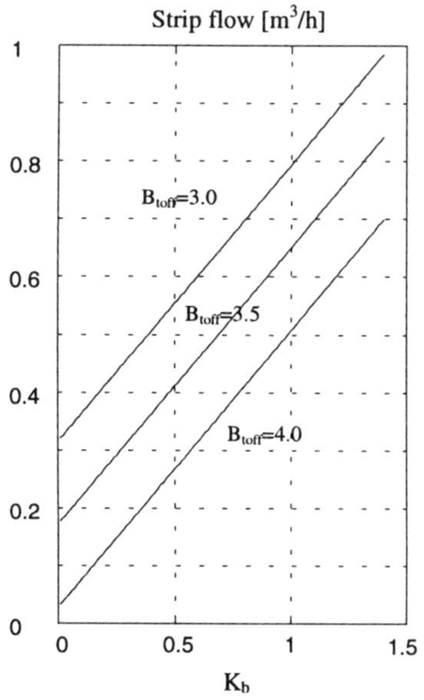

Fig. 7.6 Flow via the strip.

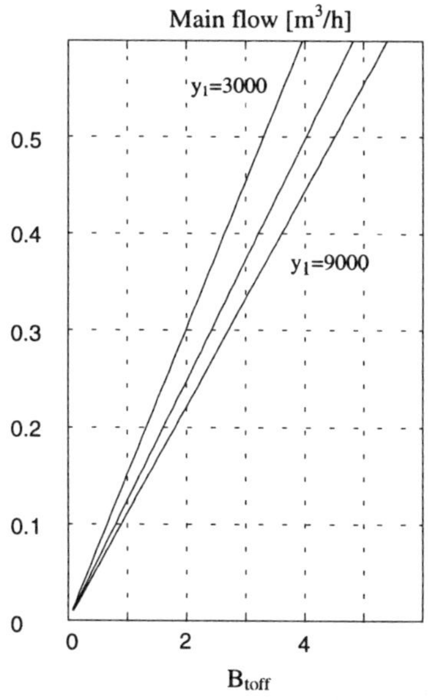

Fig. 7.7 Consumption of water.

Equation (7.46) gives the compound condition to achieve the specified conductivity value in rinse tank 5 at the stationary point. Note that the result can be given in several ways. A bad pair of squeezer rolls can be compensated for by a good pair of squeezer rolls. We shall discuss when it is necessary to change a bad pair of squeezer rolls based on the compound influence of all pairs. We have from equation (7.46):

$$\left(1+\frac{F_{c1}}{F_{b2}}\right)\left(1+\frac{F_{c1}}{F_{b3}}\right)\left(1+\frac{F_{c1}}{F_{b4}}\right)\left(1+\frac{F_{c1}}{F_{b5}}\right)=\frac{y_1}{y_5} \tag{7.58}$$

The flow via the strip is based on the estimated values of K_{bi} and B_{toffi}. Equation (7.58) gives the necessary flow of clean water fed into rinse tank 5.The values of F_{bi}, are given in table 7.4 for the combined and optimal controller. The corresponding flow of clean water computed from equation (7.58) is 1.5 [m^3/h] for the combined controller and 1.3 [m^3/h] for the optimal controller.

From the discussion in this chapter, the upper limit of the flow of clean water at the stationary point is 3.0 [m^3/h]. This means that it is not necessary to change the pair of squeezer rolls 4 during the experiment with the combined controller at once. Instead, the change can wait until the next planned stop for maintenance. The controller will still keep the conductivity rate in rinse tank 5 below the specified level, which also is shown from the results of the experiment in Chapter 6. Based on the new controllers, see Chapter 6, the new rules for replacing a pair of squeezer rolls can be summed up as follows, (Sohlberg, 1993):

- **Rule 1:** Change a pair of squeezer rolls at the next planned stop for maintenance when the relation given by equation (7.57) is not fulfilled. This rule means that the individual efficiencies of the squeezer rolls are considered.

- **Rule 2:** Change the worst pair of squeezer rolls at once when the compound influence of the squeezer rolls needs a flow of clean water larger than 3.0 [m^3/h]. The flow of clean water is computed from equation (7.58). This rule includes the influence of all pairs of squeezer rolls. However, a bad pair is compensated for by a more efficient pair. The individual limit from rule 1 can, therefore, be exceeded.

7.4.4 Economic aspects of changing the squeezer rolls

Normally, the process is stopped every fortnight for maintenance. During these stops the two double pairs of squeezer rolls and at least one of the other pairs are replaced. It is expensive to change the rolls; we will, therefore, discuss in what way some money can be saved in the changing of rolls.

The clean water is a secondary product from the heating of the rinse water. The heating is done using a heat exchanger and the clean water is obtained when the steam from the heat exchanger is cooled down. Consequently, there is no extra cost for clean water up to a consumption of 3.0 [m^3/h]. The waste water from the rinsing process is used in another process.

The costs of retreading a roll are about 1500 US$ and it costs about 100 US$ to grind a roll. Statistical data from SSAB indicate that 20% of the replaced rolls are retreaded and 80% are ground. The total cost per year of renovating the squeezer rolls is given by equation (7.59).

$$E_{tot} = \frac{w \cdot nr}{nw}\left[0.2 \cdot E_{retread} + 0.8 \cdot E_{grind}\right] \qquad (7.59)$$

$w=48$	Number of weeks per year when the process is operating
$n_r=10$	Number of rolls replaced during a stop for maintenance
n_w	Number of weeks between the stops for maintenance
$E_{retread}$	Cost of retreading one roll
E_{grind}	Cost of grinding one roll

From the experiment with controlling the process, described in Chapter 6, we have seen that the consumption of clean water is less than 50% of the water supplied. This means that by supervision and control of the process it is possible to use the rolls longer, and thus to prolong the periods between these stops for maintenance. When we increase the number of weeks between the stops, the cost per year of renovating the squeezer rolls is reduced, see Fig. 7.8.

The upper limit of the flow of clean water into tank 5 at the stationary point is 3.0 [m^3/h]. For a renovated pair of squeezer rolls, it is reasonable to assume that the flow via the strip is about 0.1 [m^3/h]. With these flows, the sum of the logarithm of the reductions is estimated to be 13.4. This is the best reduction we can obtain. When the conductivity in tank 1 is about 6000 [mS/m] and that value in tank 5 is about 0.9 [mS/m] we have from equation (7.49) that the reduction should at least be 8.9.

For the squeezer rolls used during the experiments with the new controllers, the sum of the logarithm of the reduction was between 11.0 and 12.0, based on a flow into tank 5 of 3.0 [m^3/h]. From this discussion we can estimate that the squeezer rolls should be replaced when the compound efficiency of the squeezer rolls is reduced to half of the limit for a well-rinsed steel strip. Consequently, the time between the planned stops can be prolonged. The time can be increased from two to three weeks, at least.

When the number of weeks between the stops is increased from two to three weeks, we can see from Fig. 7.8 that the cost for rolls is reduced by 25 000 US$ per year.

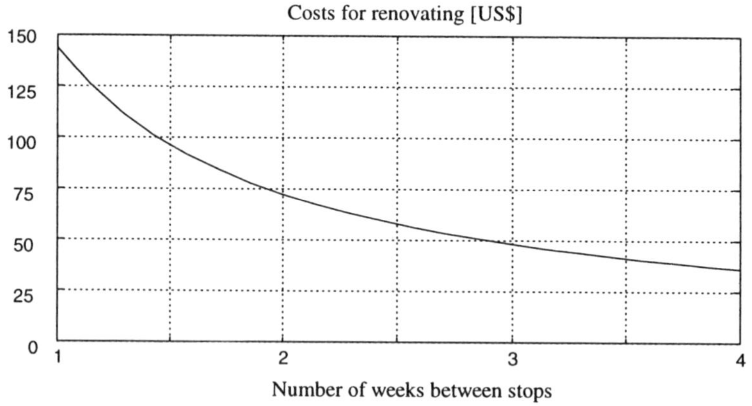

Fig. 7.8 Costs per year of renovating the rolls.

The total production volume can also be increased by reducing the number of planned stops since the pickling line is a bottle-neck. This means that the economic contribution margin can be increased and there is more money to be made.

There are similar rinsing processes in other parts of the steel works; for example, at the entry of the continuous annealing line, the steel strip passes a pickling line with a rinsing process. So there is even more money to be saved by supervision and control of those processes.

7.4.5 Evaluation of process components

The efficiency of the squeezer rolls detoriates when the process is running, the surface of the rolls becomes rougher and the flow via the top and bottom side of the strip increase. Furthermore, the hardness of the rubber increases when the rolls are used. This means that the flow besides the squeezer rolls also increases.

The rubber can be applied on the iron rolls in different ways and it is also possible to use different rubber material. This will influence the performance and detoriation of the squeezer rolls. To decide which sort of rolls should be used, three different manufacturers of squeezer rolls each delivered two sorts of rolls. Here the manufacturers are named; A, B and C. Their sort of rolls are called 1 and 2 respectively. This means that in all six different sort of squeezer rolls are subjected to testing.

The rolls were tested for about ten weeks. During the test, the unknown parameter vector $\theta(k)$ was estimated. The estimated parameters K_{bi} are used as a measure of the efficiency of the rolls to reduce the flow via the top and bottom side of the strip and the parameter B_{toffi} is used as a measure of the efficiency in reducing the flow beside the strip.

The parameters, K_{bi}, vary between 0.2 and 0.8, and the parameter B_{toffi} varies between 2.5 and 5.5. Since the production of clean water is limited to 3.0 [m^3/h], there is a limit when the main flow of clean water into zone 5 cannot compensate for the flow via the strip.

The flow beside the strip due to variations of strip thickness can with certainty be compensated for by controlling the flow of clean water into tank 5 when the parameter B_{toffi} is larger than 4.0. This gives that the parameter K_{bi} should be less than 0.5, because the production of clean water is limited. From this discussion, the performance of the squeezer rolls is satisfactory when K_{bi} is less than 0.5 and $Btoff_i$ is larger than 4.0.

The results are shown in table 7.5, (Sohlberg, 1995), where:

Kn: This sort of rolls does not meet the demand on parameter K_{bi}.

Ky: This sort of rolls meets the demand on parameter K_{bi}.

Bn: This sort of rolls does not meet the demand on the parameter B_{toffi}.

By: This sort of rolls meets the demand on the parameter $Btoff_i$.

Table 7.5 Result of the evaluation.

	A	**B**	**C**
1	Ky/By	Kn/By	Ky/Bn
2	Ky/By	Kn/Bn	Ky/By

❒ **Comments on table 7.5.** Both sorts of squeezer rolls from manufacturer A and sort 2 from manufacturer C meet the demand on both parameters K_{bi} and B_{toffi}. This means that these sorts can be recommended for use in the process. The other sorts should not be used.

7.5 Summary

In this chapter, we formed a framework for a model-based supervision system. The model is used for on-line estimation of time varying parameters. The supervision system is applied to a steel strip rinsing process and used as an advisory system to guide the process operator when parts subject to wear should be replaced.

For the specific process, a basis is developed for making a decision on when a part subject to wear should be replaced. The reduction in conductivity between adjacent rinse tanks is a measure of the efficiency of the squeezer rolls between the rinse tanks. The reduction is related to a stationary point, based on the estimates of the unknown parameters and a combination of production parameters. By comparing the reduction of the squeezer rolls, we can decide which pair is most efficient and which is least efficient. This means that we have information about which pair, if any, is in line to be replaced at the next planned stop for maintenance. In this way, we do not unnecessarily replace a well-performing pair of squeezer rolls.

We also give new rules for replacing a pair of squeezer rolls. These rules are based on the assumption that the process is controlled by a controller which involves the condition of the process. Finally, we discuss the economic aspects of predictive maintenance. By supervision and control of the process the periods between the planned stops can be prolonged.

8 FAULT DETECTION, DIAGNOSIS AND IDENTIFICATION

8.1 Introduction

In this work we distinguish between two kind of faults: slowly changing and abrupt changing faults. The first often originate from wearing and the latter from a abrupt errors within the process control system. In practice, a supervision system has to deal simultaneously with both types of faults, but here we make a distinction because different methods have to be used.

In this chapter, fast changing faults are treated and we divide the procedure into three different phases:

(i) *Fault detection.* This means discovering that a fault has occurred. It is often a matter of probability, since fault detection is based on previously collected measurements. These data are corrupted by disturbances, which influence the possibility of making judgements about a "true" fault and a "false" fault.

(ii) *Fault diagnosis.* Determine the location and type of fault occurs. It is a matter of finding out in which part of the process a fault has originated and if possible of saying more precisely which process component has failed. It also means estimating the size of the fault.

(iii) *Fault identification.* A decision is made whether it is possible to proceed with running the process in spite of the fault. Normally, the stops for maintenance are scheduled for a whole year. These stops are co-ordinated with stops for other processes. An unplanned stop causes production effects on adjacent processes and gives rise to loss in the production.

There are several survey papers in the area of fault analysis of abrupt changes in industrial systems. An example focused on fault detection methods is presented by Willsky (1976). The methods include design of failure sensitive filters and use of statistical test of the innovation sequences. An overview paper which covers fault monitoring and diagnosis in mining equipment is written by Sottile and Holloway (1994). Characteristic of this kind of process is the presence of an atmosphere which is damp, dirty and potentially explosive. The authors conclude with a comparative discussion summarising the advantages and disadvantages of the method presented.

Fault tolerance in process control is treated by Cuyvers et al. (1990). This paper presents the possibilities and limitations of different hardware fault tolerant techniques and comments on software tolerance, multiprocessor and distributed systems. A comprehensive treatment of the subject is given in a book edited by Patton et al. (1989).

In Section 8.2, we give a presentation of useful methods in the field of fault detection, diagnosis and identification. Statistical methods are given by the Weighted Sum of Squared Residuals, (WSSR). This method is complemented with the General Likelihood Ratio test, (GLR). These methods are used to detect faults with some probability. Both methods are based on the innovation process. A useful technique for failure detection is application of a bank of Kalman filters, which are used to detect abrupt faults. The method is also used here for failure diagnosis. Section 8.2 is finalised with failure identification.

In Section 8.3, we apply the WSSR-method to detect process faults in connection with supervision of wear within the process. The degree of wear is estimated by using an Augmented Extended Kalman Filter.

8.2 Fault-related methods

8.2.1 Fault detection

Faults and disturbances act in the actuators, the process and the sensors. Model-based fault detection for industrial processes means discovering the malfunction from measurements corrupted by disturbances.

Before we use more sophisticated fault detection methods, it is a good idea to check the values of the measurements. By incorporating the knowledge about the size and transient limits, it may easily be possible to detect faults. Another simple method is to plot the mean and variance of residuals.

Weighted Sum Squared Residual

The weighted sum squared residual, (Willsky 1976) and (Tzafestas and Watanbe 1990), is one of the simplest methods to detect faults. It is based on the residual sequence generated as:

$$e(k) = y(k) - \hat{y}(k) \qquad (8.1)$$

where y(k) is the process output vector and $\hat{y}(k)$ is the predicted output vector. When the process is operating normally, the innovation process is a zero mean white noise process with the covariance matrix R(k). A quantity named Weighted Sum of Squared Residual (WSSR) is defined as:

$$\mathcal{N}_N = \sum_{j=k-N+1}^{k} e(k)^T R(k)^{-1} e(k) \qquad (8.2)$$

which is a chi-squared random variable with $N \cdot p$ degrees of freedom, where p=dim[y(k)]. If a fault occurs, the statistical properties of e(k) are changed and the detection rule for failure is formulated:

$$\mathcal{N}_N(k) = \begin{cases} > \varepsilon \Rightarrow \text{Fault} \\ \leq \varepsilon \Rightarrow \text{No Fault} \end{cases} \qquad (8.3)$$

With a chi-squared table, it is possible to decide the probability that a fault has occurred, which is related to the length of the innovation window and the decision threshold ε. The window length is a trade-off between the probability of false and missed alarm. As the length of the window increases, the probability of correct detection decreases because the effect of a fault will be smoothed out.

Generalised Likelihood Ratio Technique

The method is based on the model describing the normal behaviour of the system and given in a linearized form:

$$x(k+1) = \Phi(k)x(k) + \Gamma(k)u(k) + v(k) \tag{8.4}$$

$$y(k) = Cx(k) + w(k) \tag{8.5}$$

Sudden faults result in changes as jumps and steps in the states and the output signal. The model given by equation (8.4) and (8.5) is expanded to include these possible faults:

$$x(k+1) = \Phi(k)x(k) + \Gamma(k)u(k) + v(k) + f(k)\gamma \tag{8.6}$$

$$y(k) = Cx(k) + w(k) + g(k)\gamma \tag{8.7}$$

where f(k) and g(k) are vectors of possible fault influence. The parameter γ is the magnitude of the fault. The vectors f(k) and g(k) are of the type:

$f(k)=[0\ 0\ \dots\ 1\ \dots\ 0\ 0]^T$ and $g(k)= [0\ 0\ \dots\ 1\ \dots\ 0\ 0]^T$.

where the ones, select a possible fault acting on a state or output. The vectors are equal to zero before any fault has occurred. Now, suppose that a fault of size γ has occurred at time τ; we can formulate, (Tzafestas and Watanbe, 1990):

$$x(k) = \overline{x}(k) + \alpha(k,\tau) \tag{8.8}$$

$$\hat{x}(k+1) = \hat{\overline{x}}(k) + \mu(k,\tau)\gamma \tag{8.9}$$

$$e(k) = \overline{e}(k) + \rho(k,\tau)\gamma \tag{8.10}$$

$$y(k) = C\hat{x}(k) + \beta(k,\tau)\gamma \tag{8.11}$$

where $\overline{x}(k)$, $\hat{\overline{x}}$ and $\overline{e}(k)$ are the results based on no fault. The parameters $\alpha(k)$, $\beta(k)$, $\mu(k)$ and $\rho(k)$ are calculated from the following equations:

$$\alpha(k+1,\tau) = \Phi(k)\alpha(k,\tau) + f(k,\tau) \tag{8.12}$$

$$\begin{aligned}\beta(k+1,\tau) = {} & [I - K(k)C(k+1)\mu(k+1,\tau) \\ & + K(k+1)[C(k+1)\alpha(k+1),\tau) + g(k+1,\tau)]\end{aligned} \tag{8.13}$$

$$\mu(k+1,\tau) = \Phi(k+1)\beta(k,\tau) \tag{8.14}$$

$$\rho(k,\tau) = C(k+1)[\alpha(k,\tau) - \mu(k,\tau)] + g(k,\tau) \tag{8.15}$$

We formulate two hypotheses to form a measure to detect fault. The first means no fault has occurred and the second means a fault is disturbing the system.

$\mathcal{H}_0$	E[e(k)]=0	Cov[e(k)]=R(k)
$\mathcal{H}_1$	E[e(k)]=ρ(k,τ)γ	Cov[e(k)]=R(k)

Note, the two hypotheses mean that the mean value of the innovation process is different but the covariance matrices are equal.

The likelihood function is obtained by the following expression:

$$\mathcal{L}(k,\tau,\gamma) = -\frac{1}{2}\sum_{j=\tau}^{k}[e(j) - \rho(j,\tau)\gamma]^T R^{-1}(k)[e(j) - \rho(j,\tau)\gamma] + \frac{p}{2}\mathrm{Log}\,\det(R(k)) \tag{8.16}$$

The parameter includes also the natural logarithm of the determinant of the covariance matrix R(k). The likelihood estimate with respect to γ is achieved by:

$$\hat{\gamma}(k) = \frac{\psi(k,\tau)}{\alpha(k,\tau)} \tag{8.17}$$

$$\psi(k,\tau) = \sum_{j=\tau}^{k}\rho^T(j,\tau)R^{-1}(j)e(j,\tau) \tag{8.18}$$

$$\alpha(k,\tau) = \sum_{j=\tau}^{k}\rho^T(j,\tau)R^{-1}(j)\rho(j,\tau) \tag{8.19}$$

The equation (8.17) gives the likelihood function:

$$\mathcal{L}(k,\tau,\hat{\gamma}) = -\frac{1}{2}\sum_{j=\tau}^{k} e^{T}(j)R^{-1}(j)e(j) + \frac{1}{2}\frac{\psi^{2}(k,\tau)}{\alpha(k,\tau)} \tag{8.20}$$

In equation (8.20), the summation term describes a weighted sum of squared residuals originating from the system based on hypothesis $\mathcal{H}_0$, which means no fault. The second term is a result from a fault having occurred. By calculating this quota, we have a measure for the probability of a fault. Then, the rule for detecting a fault is formulated as:

$$\frac{\psi^{2}(k,\tau)}{\alpha(k,\tau)} = \begin{cases} > \varepsilon \Rightarrow \text{Fault} \\ \leq \varepsilon \Rightarrow \text{No fault} \end{cases} \tag{8.21}$$

where ε is chosen as a compromise between false and missed alarm.

8.2.2 Fault diagnosis

Fault diagnosis is a matter of determining the location, type and size of a fault. It follows directly after the fault detection procedure. Using a simple detection method may not be enough, (Isermann, 1993). Different methods have advantages and disadvantages. For example, when we estimate the model parameters on-line and a fault occurs, the fault can force the estimates to achieve wrong values. This will mislead the process operator about the condition of the system and he might make a wrong decision about maintenance. In that a case, it is meaningful to complement the on-line parameter estimation with a fault detection method as for example the GLR or WSSR methods.

In the process industry, we have many processes which are non-linear and are subjected to wear. It might be difficult to achieve robustness of an on-line parameter estimation system for supervision of the wear. If we try to expand the estimation system with a bank of Kalman Filters to detect faults, the estimates may diverge. A solution to this problem can be to use the "back-tracking" method. It originates from the field of computer programming and means a decision is taken based on unsure information. On such an occasion, the computer program is constructed to proceed based on a guess which path is the right one. Later, if the program discovers from other information a wrong decision was taken, a new guess is made leading to another path.

For an industrial process, suppose a fault has occurred at time τ, where $k > \tau$, and we apply the method of back-tracking diagnosing a failure. For this type of application we have to store the input-output data and estimate wearing parameter from time τ. From that moment, we start to run a bank of filters based on fixed values of parameters estimating the wear. This is possible due to the fact that the wear is a slowly varying process.

The bank of Kalman filters is based on a priori knowledge about which kind of faults may occur. This is a weakness in the method, since it is very difficult to know all types of possible faults. However, when a known fault occurs, the type and the size of the fault is estimated with good precision. A schematic outline of fault diagnosis by using a bank of Kalman filters is shown in Fig. 8.1. The data base block communicates with the block of parameter estimates and stores information about inputs, outputs and parameter estimates, which are used by the bank of filters for fault analysis.

Another model-based fault diagnosis method is to make use of the innovation process. For many multi-output systems, the residuals are related to a definite part of the process and make it possible to locate the fault.

Model-based failure diagnosis can also be based on a causal relationship between fault and symptoms, (Isermann, 1997). Causality is followed by a tree analysis giving information about the location and what kind of fault has occurred. In practice, the causalities are not known exactly, so when using this kind of method, we will have an on-line engineering expert system. It is a combination of a formal engineering subject and an expert system. The system will partly be based on human reasoning and formal rules.

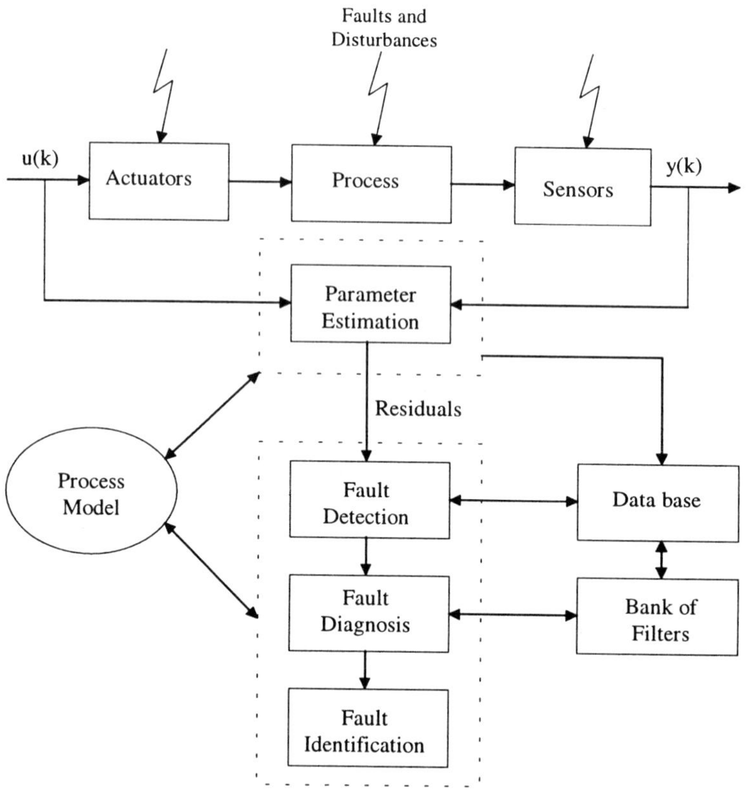

Fig. 8.1 Fault diagnosis based on a bank of Kalman filters.

8.2.3 Fault identification

When a fault is detected and diagnosed, a decision must be taken whether it is possible to proceed with running the process or not. We have to investigate the influence on the resulting product quality. Furthermore, we have to compare the costs for running the process marred with faults with the costs for stopping the process and taking care of the fault.

By incorporating the kind and size of a fault into the model, the behaviour of the process can be analysed by simulation and an operator can be warned of imminent undesirable states that the process might enter, (Dvorak and Kuipers, 1991). Similarly, the effects of proposed control actions can be determined and analysed by simulation. We can also analyse if it is possible to compensate for the fault by an controller, which includes the changed behaviour of the process.

An easy way to design fault identification, is to use the statistical properties of the process, like the magnitude, the mean and the variance of important variables and check the values. The fault identification procedure should also include a module for risk analysis. It might be possible to run a process with a fault, but the risk of a serious breakdown can be increased considerably. For example, when one looper of a rolling mill is out of order, it is still possible to run the mill, but when another looper is damaged, the strip will break and the process will be closed for a long time.

8.3 Application of fault-related methods

A wearing supervision system presented in Chapter 7. The system is based on the steel strip being completely rinsed after passing each rinse zone. This means that the concentration of the liquid on the strip after a pair of squeezer rolls is the same as in the preceding rinse tank.

However, there are situations when the strip will not be completely rinsed after passing a rinse zone. This is due to faults, for example:

- The rinse nozzles are blocked.
- The rinse nozzles are damaged in some way and the rinse water is not sprayed in a proper direction on the steel strip.
- The level of the rinse water in some of the tanks is too low.
- The pump system for the circulated rinse flow is not working properly.

When the strip is not completely rinsed, the concentration of the liquid on the strip after a pair of squeezer rolls is higher than the concentration in the rinse tank which the strip has passed. Since the estimates of the parameter vector $\theta(k)$ are based on a well-rinsed strip, the parameter vector will be biased. This means the advisory system will mislead the process operators about the wearing of squeezer rolls.

❒ **Simulation with fault.**

As the "process" during the simulation, we use a model of the process where all unknown parameters in the process model are identified off-line by optimising the likelihood function. The identification is made with a measured data sequence collected from the real process. The simulation is made with the corresponding sequence of production parameters. The process is affected by the combined controller presented in Chapter 6.

The estimation of the unknown parameters, describing the wear, is made by the same Extended Kalman Filter as used in Chapter 5. The filter is also tuned with the same values as used in Chapter 5.

A fault in the process is generated after 20 hours simulation. The purpose of the fault is to imitate the situation when some of the nozzles in zone 3 are blocked. This means that the steel strip will not be completely rinsed when strip velocity is above a certain limit, B_{vlimit}. After rinse zone 3, the flow via the strip will contain one part rinse water from tank 3 and one part rinse water from tank 2.

When a fault has occurred, the concentration of the flow via the strip after tank 3 is formulated as:

$$C_{strip} = (1-\beta)\cdot C_3 + \beta \cdot C_2 \qquad (8.22)$$

$$\text{where } \beta = \begin{cases} \dfrac{B_v - B_{v\,limit}}{B_{v\,max}} & B_v \geq B_{v\,limit} \\ 0 & B_v < B_{v\,limit} \end{cases}$$

Equation (8.22) means that the concentration of the flow via the strip fed into tank 4 is greater than the concentration in tank 3 when the strip velocity is above the velocity limit. During the simulations, the velocity limit is set to 125 [m/min.]

Comments on the parameter estimates.

The parameter estimates are shown in Fig. 8.2 and Fig. 8.3. From the simulation, it is seen that the estimate of parameter K_{b4} is increased after 40 hours, due to a higher concentration of the flow via the strip. Based on only the estimated parameter, one might believe that the efficiency of the squeezer rolls 4 has decreased. However, it is known from fact that this is not the case. Instead some of the nozzles in rinse zone 3 are blocked. Consequently, a supervision system using only the estimates from wearing may mislead the operator.

Comments on fault detection, diagnosis and identification.

Fault detection. The weighted sum squared residual technique is used to detect a fault. In our case we chose that the length of the window should be N=10. The output vector is p=5. The threshold for a failure is given from a chi-2 table as 75 when $N \cdot p=50$ degrees of freedom and 0.99 confidence level.

The high value of the WSSR after 40 hours indicates that a fault has occurred, see Fig.8.4. As can be seen from Fig. 8.4, the subsequent peaks of the WSSR are decreased. This is due to the unknown parameters being adapted to the new situation with a failure.

Other simulations show that the parameter B_{vlimit} has to be lower than 150 [m/minute] to detect a failure with 0.99 confidence level. Simulation with wear of squeezer rolls 4 and where the parameter K_{b4} is doubled during the simulated period, does not give any significant result of the WSSR. This study is not presented here. Consequently, normal wear of the squeezer rolls will not be detected as a failure by the WSSR-method. It is also shown by simulation that a parameter has to be altered more than 100% to detect an abrupt failure of a pair of squeezer rolls by the WSSR method. This means that the squeezer rolls have to be seriously damaged.

Fault diagnosis. When a failure is detected with the WSSR technique, the fault can be isolated to the failed rinse zone by using the respective residual sequence. From the simulated case, the residual from each respective rinse zone is shown in Fig. 8.5. When the failure in rinse zone 3 occurs, the residual sequence from zone 4 is noticeably increased compared to the other residual sequences.

Fault identification. The concentration of acid will increase when an abrupt failure occurs or the squeezer rolls wear out. The controller can compensate for this situation so that the concentration in tank 5 will be below the specified limit to achieve a well-rinsed strip. The mean value of the flow into tank 5 is calculated and used to decide whether the process should be stopped when an abrupt failure occurs. The mean value is given by equation (8.23).

$$u_N(k) = \frac{1}{N} \sum_{i=k-N+1}^{k} u(k) \qquad (8.23)$$

The window N must not be too short because $u_N(k)$ will be too sensitive to peaks in the control signal u(k). Reverse, when the window is too long, $u_N(k)$ will be insensitive to variations in u(k). The condition required to stop the process and take care of a failure is fulfilled when the mean value $u_N(k)$ given by equation (8.23) reaches the limit of produced clean water.

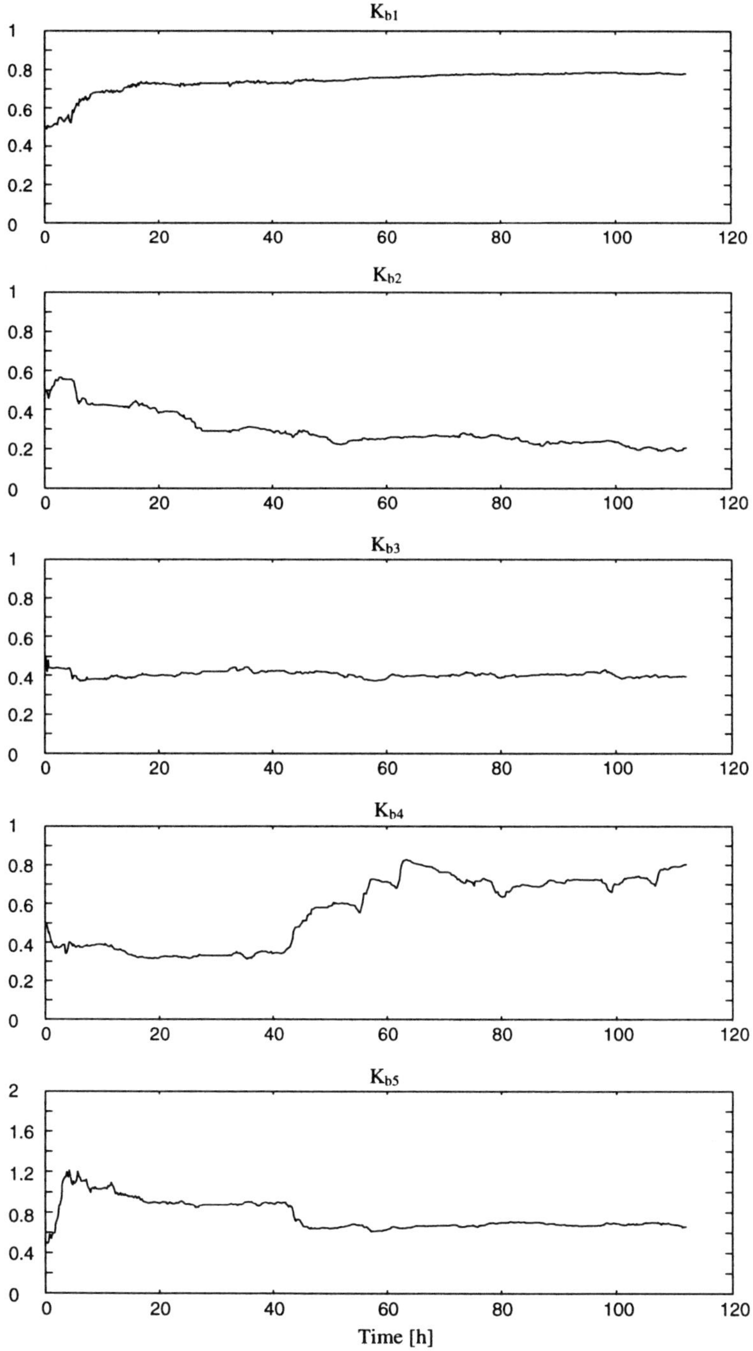

Fig. 8.2 Estimated parameters in the case of simulated fault detection.

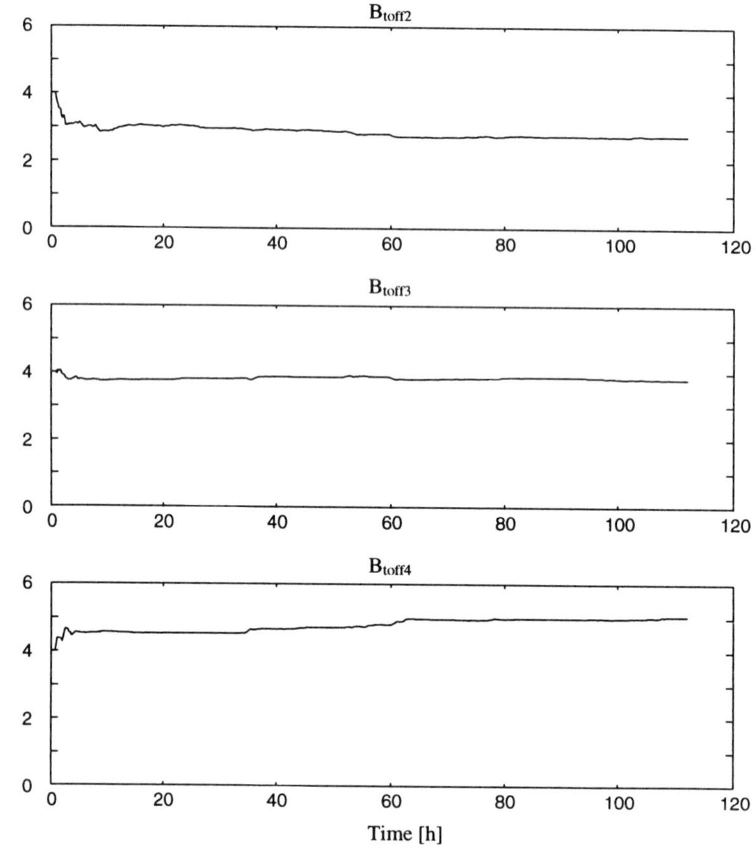

Fig. 8.3 Estimated parameters in the case of simulated fault detection.

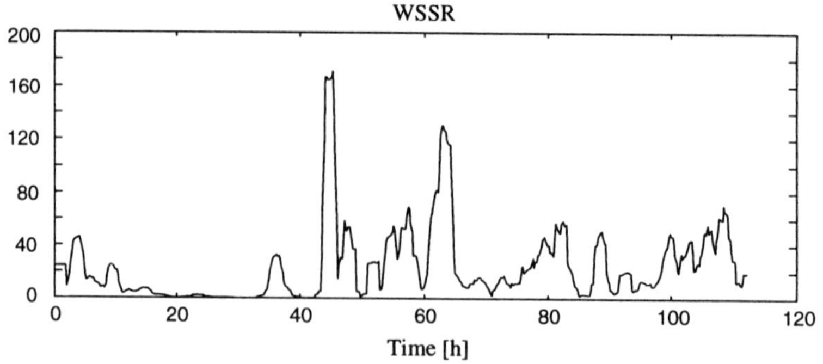

Fig. 8.4 Weighted sum of squared residuals in the case of simulated fault detection.

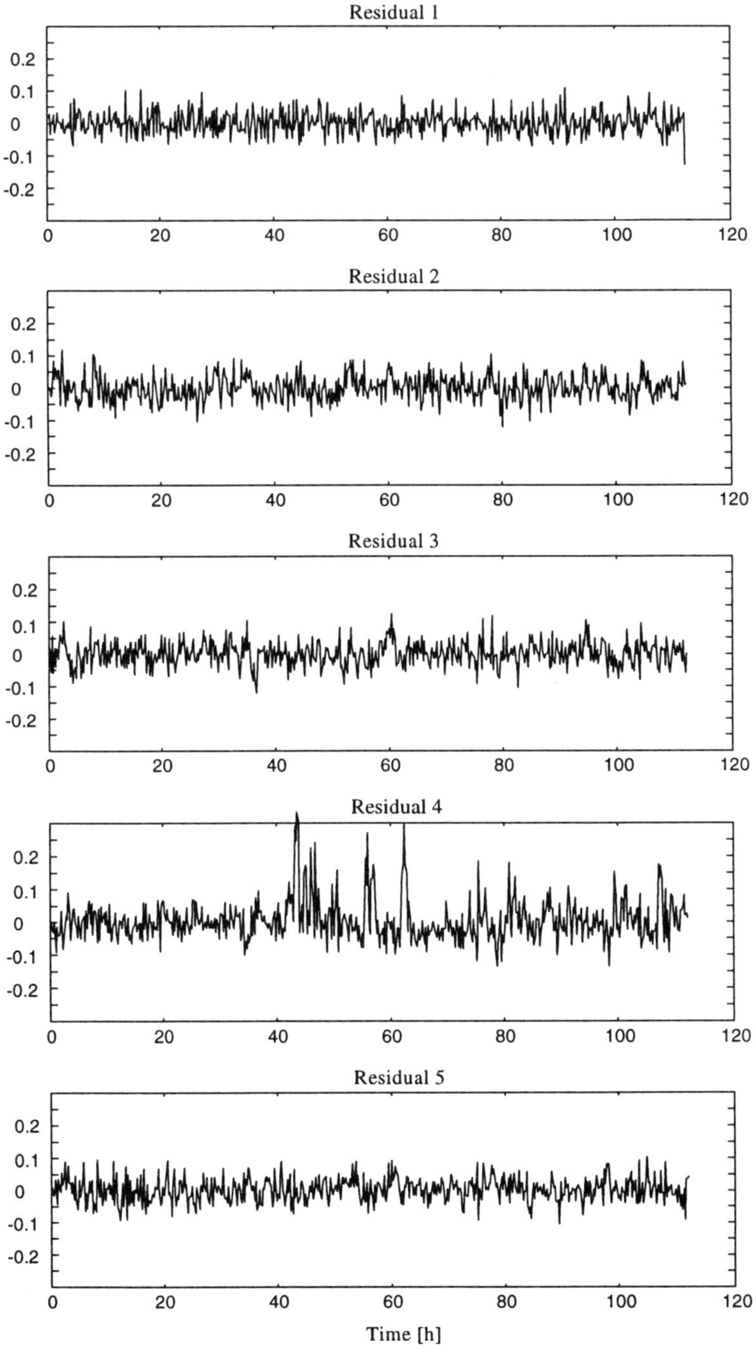

Fig. 8.5 Residuals from tank 1 to tank 5 in the case of simulated fault detection.

8.4 Summary

This chapter deals with abrupt faults and considers fault detection, diagnosis and identification. The weighted sum squared residuals is a relatively simple method to implement but a decision threshold must be calculated for proper functioning. Another method presented is the generalised likelihood ratio technique. It is somewhat more complicated to implement but the decision threshold is chosen as a compromise between false and missed alarm.

We have also presented two methods of using a bank of Kalman filters to detect fault. The advantage of the methods is the possibility to diagnosis the fault and control the process by using a filter which takes care of the fault. The disadvantage is that the bank of filters can be huge when all probable faults are covered. A special reduced order bank of Kalman filters is also discussed. This method is specially suited for detection of sensor faults.

Fault diagnosis is sometimes integrated with the fault detection methods, as when using a bank of Kalman filters. Another method is to make use of the innovation process. The residuals are often related to a definite part of the process. Fault identification is related to economic and quality aspects of running the process. The question is whether the process should be stopped when a fault occurs or if it is possible to proceed with running the faulty process.

The industrial process which is studied in this work has parts which wear out when the process is running. An Extended Kalman Filter is used to estimate both the process states and the unknown parameters describing the parts subject to wear. A supervision system based only on parameter estimation may give incorrect information about the amount of wear of the worn parts, when a fault occurs apart from normal wear.

Therefore, the supervision system presented in Chapter 7 is complemented with a fault detection system, based on the weighted sum of squared residuals. When the process is running during normal operation, data from the process will give "normal" statistical properties. To decide if a failure has happened, we analyse with the WSSR technique whether the statistical properties have changed significantly. The fault is also located to a definite part of the process and we analyse if it is possible to go on with running the process without immediately repairing of the failure.

REFERENCES

Akiake H. (1981). *Modern development of statistical methods. Trends and Progress in System Identification.* Pergamon Press. Oxford.

Anderson B.D.O. and Moore J.B. (1979). *Optimal Filtering.* Prentice-Hall. Englewood Cliffs, New Jersey.

Atkinsson A. C. (1975). *Planning Experiment for Model Testing and Discrimination.* Math. Operationsforsch. u. Statist. 6, Heft 2. pp. 253-267

Balchen J.G., Ljunquist D. and Strand S. (1992). *State Space Predictive Control.* Modelling, Identification and Control. Vol. 13, No. 2 pp. 77-112. Norwegian Research Bulletin, Oslo

Baumann (1991). *Discrete-time control of Continuous-time Non-linear Systems.* International Journal of Control. Vol. 53, No. 53, pp. 113-128.

Bellman R. E. (1957). *Dynamic Programming.* Princeton, New-York: Princeton University Press.

Bellman R. E. and Kalaba R. E. (1965) *Dynamic Programming and Modern Control Theory.* New-York. Academic Press.

Bohlin T. (1970) *On the Maximum Likelihood method of Identification.* IBM Journal Research Development no 14, pp. 41-51 Stockholm.

Bohlin T. (1984) *Computer Aided Grey-Box Identification.* TRITA-REG 8403. Royal Institute of Technology. Stockholm

Bohlin T. (1991a). *Interactive System Identifications: Prospects and Pitfalls.* Springer-Verlag, Berlin.

Bohlin T. (1991b). *Grey-box Identification: A Case Study.* TRITA-REG 91/0001. Royal Institute of Technology, Stockholm Sweden.

Bohlin T. (1994) *A Case Study of Grey Box Identification..* Automatica Vol 30 No 2. pp. 307-318. Pergamon Press. London.

Box G.E.P. and Tiao G.C. (1973). *Bayesian Inference in Statistical Analysis,* Addison-Wesley, Reading, Massachusetts.

Camacho E. F. and Bordons C. (1995). *Model Predictive Control in the Process Industry.* Advances in Industrial Control. Springer-Verlag.

Camo A/S (1987). *Modelling & Identification.* Trondheim Norway

Chung K. L. (1979). *Elementary Probability Theory with Stochastic Processes.* 3rd edition. Springer-Verlag New York.

Cuyvers R., Lauwereins R. and Peperstraete J.A. (1990). *Fault Tolerance in Process Control: possibilities, limitations and trends.* Journal A. Vol. 31. No. 4. pp. 33-40.

Dvorak D. and Kuipers B. (1991). *Process Monitoring and Diagnosis, A Model-Based Approach.* IEEE Expert/Intelligent Systems. June pp. 67-74.

Eykhoff P. (1974). *System Identification, Parameter and State estimation.* Wiley & Sons.

Goodwin G. C. (1971). *Optimal Input Signals for Nonlinear System Identification.* PROC. IEE, Vol. 118. pp. 922-926

Goodwin G. C. (1982). *Experiment Design. IFAC Symposium on System Parameter Estimation.* Washington DC. USA.

Goodwin G.C. and Payne R.L. (1977). *Dynamic System Identification: Experiment Design and Data Analysis.* Academic Press. New York.

Grace A. (1990). *Optimisation Toolbox, Matlab.* The MathWorks Inc. USA

Graebe S. (1990a). *IDKIT. A Software for Gray Box Identification. Mathematical Reference*. Royal Institute of Technology. TRITA-REG 90/00003. ISSN 0347-1071. Stockholm

Graebe S. (1990b). *IDKIT. A Software for Gray Box Identification. User Guide.* Royal Institute of Technology. TRITA-REG 90/00004. ISSN 0347-1071. Stockholm

Graebe S. (1990c). *IDKIT. A Software for Gray Box Identification. Implementation and Design Reference.* TRITA-REG 90/00005. ISSN 0347-1071. Stockholm

Grewal M.S. and Andrews A.P. (1993). *Kalman Filtering: Theory and Practice.* Prentice-Hall. Englewood Cliffs, New Jersey.

Handbook of Engineering. (Ingenjörshandboken in Swedish)), N.R. Stockholm, Sweden.

Harvey A. C. (1989). *Forecasting Structural Time Series Models and the Kalman Filter.* Cambridge University Press.

Henson M. A. and Seborg D. E. (1997). *Nonlinear Process Control.* Prentice Hall.

Holst J. Holst U. Madsen H and Melgaard H. (1992). *Validation of Grey Box Models.* IFAC Adaptive Systems in Control and Signal Processing, Grenoble, France. pp. 53 -60.

Isermann R. (1984). *Process Fault Detection Based on Modelling and Estimation Methods - A Survey.* Automatica. Vol. 20, No. 4, pp. 387-404.

Isermann R. (1989). *Process Fault Diagnosis Based on Dynamic Models and Parameter Estimation Methods.* pp 253-291. Fault Diagnosis in Dynamic Systems, Theory and Applications. Edited by R. Patton, P Frank and R. Clark. Prentice Hall.

Isermann R. (1989). *Process Fault Diagnosis Based on Process Model Knowledge.* IFAC, Advance Information Processing in Automatic Control. pp. 21- 34. Nancy, France.

Isermann R. (1993). *On the Applicability of Model Based Fault Detection for Technical Processes.* IFAC World Congress. Sydney.

Isermann R. and Balle P. (1996). *Trends in the Application of Model Based Fault Detection and Diagnosis of Technical Processes.* pp. 7f-01 1-12. 13th Triennial World Congress, San Francisco. USA.

Isermann R. Lachmann K.-H. and Matco D. (1992). *Adaptive Control Systems.* Prentice Hall International Series in Systems and Control Engineering. Prentice Hall.

Jazwinski A. H. (1970). *Stochastic Processes and Filtering Theory.* Academic Press, New-York.

Johansson R. (1993). *System Modelling & Identification.* Prentice Hall Information and System Science Series.

Kailath T. (1980). *Linear Systems.* Prentice-Hall Inc.

Kalman R. E. (1960). *On the General Theory of Control Systems.* Proceedings First IFAC Congress. pp. 481-493.

Karplus W. (1976). *The Spectrum of Mathematical Modelling and System Simulation. Simulation of Systems.* North-Holland Publishing Company.

Kato T., Sekine H., Kikuchi T. Hayasi Y. and Kuraishi T. (1992). *Development of High-Accuracy Control System for Hot Strip Mill.* Automation in Minimg, Mineral and Metal Processing. pp. 151-156.

Kematron A/S (1990). *Manual of Conductivity Measurement.* Lyngby, Denmark

Kirk D. E. (1970). *Optimal Control Theory.* Prentice-Hall Inc.

Kushner J. B. (1976). *Water and Waste Control for the Plating Shop.* Gardner Publications Inc. USA

Kwakernaak H. and Sivan R. (1972). *Linear Optimal Control System.* John Wiley & Sons Inc.

La Cava M., C Picarrdi and F. Ranieri (1989). *Application of the Extended Kalman Filter to Parameter and State Estimation of Induction Motors.* International Journal of Modelling and Simulation, Vol. 9, No.3

Lawrence and Rugh (1995). *Gain Scheduling Linear Dynamic Controllers for a Nonlinear Plant.* Automatica, 31 (3) pp. 381--390.

Ledding A. E. (1986). *Estimation Chemical Concentration in Rinse Tanks.* Metal Finishing.

Leondes C. T. (1979). *Control and Dynamic Systems, Advances in Theory and Application.* Academic Press.

Ljung L. (1987). *System Identification; Theory for the User.* Prentice Hall Information and System Science Series.

Ljung L. and Glad T. (1994). *Modelling of Dynamic System,* Prentice Hall, Englewood Cliffs, NJ, 1994, ISBN 0-13-597097-0.

Ljung L. and Söderström T. (1983). *Theory and Practice of Recursive Identification.* The MIT Press Cambridge, Massachusetts, London.

Matlab (1991). *Matlab Users Guide.* The MathWorks Inc. USA.

Maybeck P.S. (1979). *Stochastic Models, Estimation and Control* 1. Academic Press. New-York.

Maybeck P.S. (1982). *Stochastic Models, Estimation and Control* 2. Academic Press. New-York.

McFahrlane D. C. and Stone P. M. (1990). *Minimum Tension In a Merchant Bar Rolling Mill Using Modern Control Techniques.* 11th Triennial World Congress. pp. 137-142. Tallinn, Estonia.

Mehra R. K. Mehra. (1974). *Optimal Input Signals for Parameter Estimation in Dynamic Systems - Survey and New Results.* IEEE Transaction of Automatic Control, Vol. AC-19, December 1974. pp. 753-767.

Melgaard, Sadegh, Madsen and Holst (1993). *Experiment Design for Grey-Box Models.* 12th IFAC World Congress, Sydney. pp II:145--148.

Mohler J. B. (1977). *Non-Equilibrium Rinsing.* Metal Finishing.

Norton H. (1970). *Sensor and Analyser Handbook.* Prentice-Hall.

Ordys A. W., Pike A. W., Johnson M. A. Katebi R. M. and Grimble M. J. (1994). *Modelling and Simulation of Power Generation Plants.* Advances in Industrial Control. Springer-Verlag.

Patton R., Frank P. and Clark R.(Editors) (1989). *Fault Diagnosis in Dynamic Systems, Theory and Applications.* Prentice Hall International Series in Systems and Control Engineering. Prentice Hall

Pontryagin L. S., Boltyanskii V. G., Gamkrelidze R. V. and Mischenko E. F. (1962). *The Mathematical Theory of Optimal Processes.* New-York. Interscience Publishers, Inc.

Powell A. (1978). *A Fast Algorithm for Nonlinear Constrained Calculations.* Lecture Notes in Mathematics. Springer-Verlag

Press S. J. (1989) *Bayesian Statistics: Principles, Models and Applications*, John Wiley.

Rao C. R. (1973). *Linear Statistical Inference and Its Applications.* John Wiley, London.

Rao M, Xia Q. and Ying Y. (1993). *Modelling and Advances for Process Industries, Applications to paper Making Processes.* Advances in Industrial Control Series. Springer-Verlag.

Rugh (1991). *Analytical Framework for Gain Scheduling.* IEEE Control Systems Magazine, 11 (1), pp. 79--84.

Sage A. P. (1987). *Bayes rule. System and Control* Encyclopaedia, Vol. 1. Pergamon Press, Oxford.

Shamma and Athens (1992*). Gain Scheduling: Potential Hazards and Possible Remedies.* IEEE Control Systems Magazine, June 1992, pp. 101--107.

Shinskey F. (1979). *Process-Control Systems.* McGraw-Hill Book Company.

Sohlberg B. (1992). *Supervision of a Steel Strip Rinsing Process.* The 31st IEEE Conference on Decision and Control. Tucson USA. pp. 2557-2561.

Sohlberg B. (1993). *Control and Supervision of a Steel Rinsing Process.* 12th World Congress IFAC. Sydney, Australia. To be presented.

Sohlberg B. (1993). *Optimal Control of a Steel Strip Rinsing Process.* The Second Conference on Control Applications. Vancouver. pp. 247 - 363.

Sohlberg B. (1995). *Evaluation of the Supervised Performance of a Steel Strip Rinsing Process.* The 8:th IFAC International Symposium on Automation in Mining, Mineral and Metal Processing. Sun City. pp. 151 -154.

Sottile J. and Holloway L.E. (1994). *An Overview of Fault Monitoring and Diagnosis in Mining Equipment.* IEEE Transaction on Industry Applications, Vol. 30, No. 5.

Spriet J. A. and Vansteenkiste G. C. (1982). *Computer Aided Modelling and Simulation.* Academic Press.

Stein B. (1988). *Recuperative rinsing; a mathematical approach.* Metal Finishing.

Stephanopoulos G. (1984). *Chemical Process Control; An Introduction to Theory and Practice.* Prentice-Hall, INC. Englewood Cliffs, New Jersey

Söderström T. and Stoica P. (1989): *System Identification*, Prentice-Hall, London.

Tzafestas S. and Watanabe K. (1990). *Modern Approaches to System/Sensor Fault Detection and Diagnosis.* Journal A: Vol. 31 No. 4. pp. 42-57.

Wilks S. S. (1962). *Mathematical Statistics.* John Wiley, London

Willsky A. S. (1976). *A Survey of Design Methods for Failure Detection in Dynamic Systems.* Automatica Vol. 12. pp. 601-611. Pergamon Press.

Willsky A.S. (1980). *Failure Detection in Dynamic Systems.* AGARD, N0.109.

Yang Y. and Lu Y. (1988). *Dynamic Model Based Optimisation Control for Reheating Furnaces.* Computers in Industry 10. pp. 11-20. North-Holland

Åström K. J. and Wittenmark B. (1997). *Computer Controlled System.* Prentice-Hall Inc.

Åström K.J. and Wittenmark B. (1995). *Adaptive Control.* Addison-Wesley Publishing Company.

INDEX

D

E

F

G

H

I

K

L

M

N

O

P

T

U

W